NANOTECHNOLOGY SCIENCE AND TECHNOLOGY

SMART NANOOBJECTS

FROM LABORATORY TO INDUSTRY

NANOTECHNOLOGY SCIENCE AND TECHNOLOGY

Additional books in this series can be found on Nova's website under the Series tab.

Additional e-books in this series can be found on Nova's website under the e-book tab.

NANOTECHNOLOGY SCIENCE AND TECHNOLOGY

SMART NANOOBJECTS

FROM LABORATORY TO INDUSTRY

KIRILL LEVINE
EDITOR

nova publishers
New York

For permission to use material from this book please contact us:
Telephone 631-231-7269; Fax 631-231-8175
Web Site: http://www.novapublishers.com

Library of Congress Cataloging-in-Publication Data

Smart nanoobjects : from laboratory to industry / editor, Kirill Levine.
pages cm
Includes bibliographical references and index.
ISBN: 978-1-62808-741-3 (hardcover)
1. Smart materials. 2. Nanostructured materials. I. Levine, Kirill. II. Title: Smart nano objects.
TA418.9.S62S5244 2013
620'.5--dc23

2013029216

Published by Nova Science Publishers, Inc. † New York

CONTENTS

PREFACE

This book covers a variety of topics related to the synthesis of smart nanoobjects. Among them are surface nanostructured metals with improved corrosion and wear resistance (Syrkov), synthesis of organo-micro cycles, with emphasis on characterization of obtained materials by infrared spectroscopy, HPLC, and NMR analysis (El-Saied), Schiff bases metal complexes and their investigation for antibacterial and antifungal activity (El-Seidy), and transition metal complexes (Al-Sawaf).

Nanotubes have always been positioned in the study of nanocomposites as interesting materiala for applications in electronics (transistors, resistors, varistors, sensors, display devices to mention just a few). This volume includes research on thin film deposition by acoustic wave atomizer (Balachova), and expholiation of graphene from liquid phase (Moshkalev).

Continued interest in conducting polymer hydrogels have stimulated research for their application as pH sensors with sensing principles based on electroactivity (Elyashevich).

Investigation of mass transfer kinetics in bifunctional matrixes is carried out with new type of materials: metal-ion exchangers (Kalinitchev).

The world of nanotechnology described in this book will excite any reader, from interested students to industrial or academic professionals.

In: Smart Nanoobjects
Editor: Kirill Levine

ISBN: 978-1-62808-741-3

Chapter 1

SYNTHESIS AND CHARACTERIZATION OF SOME SCHIFF BASES METAL COMPLEXES AND THEIR INVESTIGATION AS ANTIBACTERIAL AND ANTIFUNGAL AGENTS

Ahmed M. A. El-Seidy *
Inorganic Chemistry Department, National Research Center,
Cairo, Egypt
Chemistry Department, Faculty of Science,
The Imam Muhammad bin Saud Islamic University,
Riyadh, Saudi Arabia

ABSTRACT

New series of Fe(III), Co(II), Cu(II) and Ni(II) of the 6,6'-((1E,1'E)-(propane-1,2-diylbis(azanylylidene))bis(methanylylidene))bis(7-hydroxy-5-methoxy-2-methyl-4H-chromen-4-one), H_2L, have been synthesized and characterized by elemental analysis, IR, UV-vis, ^{1}H-NMR, mass and ESR spectra, magnetic susceptibility and molar conductivity measurements. The spectral data and magnetic measurements of the complexes indicate that, the geometries are either square planar or

* Corresponding authors at: Inorganic Chemistry Department, National Research Centre, Dokki, 12622, Cairo, Egypt. Tel.: 002-01095105802. E-mail address: ahmedmaee2@gmail.com.

octahedral. The antibacterial and antifungal activities of the compounds showed that, some of metal complexes exhibited a greater inhibitory effect than standard drug as tetracycline (bacteria), Amphotericin B (fungi), the parent ligand and the corresponding metal ion.

Keywords: Schiff base ligand, Biological activity, antibacterial, antifungal, metal complexes

INTRODUCTION

Schiff base ligands played an important role in the development of coordination chemistry and they have been employed throughout the transition metal series. Metal complexes of these ligands are characterized by their facile synthesis, the accessibility of diverse structural modifications and the availability of chiral versions. [1, 2] Transition metal complexes with oxygen and nitrogen donor Schiff bases are of particular interest [3] because of their ability to possess unusual configurations, be structurally labile and their sensitivity to molecular environments. [4] Schiff base ligands have proven to be very effective in constructing supramolecular architectures such as coordination polymers, double helixes, and triple helicates. [5] Schiff bases can accommodate different metal centres involving various coordination modes allowing successful synthesis of homo- and heterometallic complexes with varied stereochemistry. This feature is employed for modelling active sites in biological systems. [6, 7] Additionally, they have wide applications in fields such as antibacterial, antiviral, antifungal agents, [8] homogeneous or heterogeneous catalysis [9, 10] and magnetism. [11] Schiff bases are potential anticancer drugs and when administered as their metal complexes, the anticancer activity of these complexes is enhanced in comparison to the free ligand. [12]

In view of the versatile importance of Schiff bases and their complexes, I report in this article the synthesis and characterization of the ligand 6,6'-((1E,1'E)-(propane-1,2diylbis(azanylylidene))bis(methanylylidene))bis(7-hydroxy-5-methoxy-2-methyl-4H-chromen-4-one), H_2L and its Fe(III), Co(II), Cu(II) and Ni(II) complexes. The coordination behavior of the Schiff base ligand towards metal ions was investigated via IR, UV-Vis, ESR as well as conductivity and magnetic moments measurements. The antibacterial and antifungal activities of the compounds showed that, some of metal complexes exhibited a greater inhibitory effect than standard drug as tetracycline (bacteria) and amphotericin B (fungi).

EXPERIMENTAL

1.1. Material

All the reagents employed for the preparation of the ligand and its complexes were of the best grade available and used without further purification.

1.2. Synthesis of the Schiff Base Ligand H_2L

Ethanol solution (5 mL) of propane-1,2-diamine (1.1 mmole, 7.9 X 10^{-2} g) was added to a hot 70°C ethanol solution (15 mL) of 6-acetyl-7-hydroxy-5-methoxy-2-methyl-4H-chromen-4-one (2.1 mmole, 0.5 g) and reflux for 4h then the solvent was evaporated to 10 mL. After coaling the yellow precipitate was filtered off and dried under vacuum over anhydrous $CaCl_2$, Scheme 1. ^{1}H NMR (300MHz, $CHCl_3$): δ =12.08 [S, 1H, H(22)], 11.53 [S, broad, H(37)], 8.97 [S, 2H, H(46) and H(41)], 5.77 [m, 2H, H(44) and H(42)], 5.66 [S, 2H, H(45) and H(43)], 4.04 [psd, 2H, H(39) and H(38)], 3.51[S, 6H, CH_3(36) and CH_3(21)], 3.43 [m, 1H, H (40)], 2.21 [S, 3H, CH_3(34)], 2.19[S, 3H, CH_3(19)], 1.40 [d, 3H, CH_3(3)], Scheme 2.

1.3. Synthesis of the Metal Complexes

1.3.1. Preparation of Complexes 5, 7 and 9

A hot (60° C) methanol solution of the copper chloride or copper acetate was added to a hot (60° C) 1,1-dichloroethane solution of the ligand with molecular ratios 1:1 L/M (ligand /metal) (complex **5**) and 2:1 L/M (complexes **7** and **9**). Complex **5** was obtained in case of using copper chloride or copper acetate. The reaction mixture was then refluxed for a 4 h. The precipitates formed were filtered, washed with ethanol, then with diethyl ether and dried under vacuum over anhydrous $CaCl_2$.

1.3.2. Preparation of Other Complexes

Other metal complexes of the ligand H_2L were prepared by mixing a hot (60° C) methanol solution of the metal salts with the required amount of a hot (75° C) ethanol solution of the ligand to form 1:1 or 1:2 L/M (ligand /metal). The reaction mixture was then refluxed for a time depending on the transition metal salt used. The precipitates formed were filtered, washed with ethanol, then with diethyl ether and dried under vacuum over anhydrous $CaCl_2$.

1.3.3. In-Vitro Antibacterial and Antifungal Activities

The investigation of the biological activities of the newly synthesized Schiff base ligand, its metal complexes and their corresponding metal salts were carried out in the Botany Department Lab. of microbiology, Faculty of Science, El-Menoufia University. The antibacterial and antifungal activities were investigated by disc diffusion method. [13] The antibacterial activities were done using *Escherichia coli* and *Aspergillus niger*. The tested compounds were dissolved in DMSO to give 2000 ppm stock solutions. DMSO poured disc was used as negative control. The bacteria were subcultured in nutrient agar medium which was prepared using ($g.L^{-1}$ distilled water) NaCl (5 g), peptone (5 g), beef extract (3 g), agar (20 g). The fungus was subcultured in Dox´s medium which was prepared using ($g.L^{-1}$ distilled water) yeast extract (1g), sucrose (30 g), $NaNO_3$, agar (20 g), KCl (0.5 g), KH_2PO_4 (1 g), $MgSO_4.7H_2O$ (0.5 g) and trace of $FeCl_3$ $.6H_2O$. These mediums were then sterilized by autoclaving at 120 ^{0}C for 15 min. After cooling to 45 ^{0}C the medium was poured into 90 mm diameter Petri dishes and incubated at 37 ^{0}C or 28 ^{0}C, respectively. After few hours, Petri dishes were stored at 4 ^{0}C. Microorganisms were spread over each dish by using sterile bent Loop rod. The test is carried out by placing filter paper disks with a known concentration of the compounds on the surface of agar plates inoculated with a test organism. Standard antibacterial drug (tetracycline), antifungal drug (Amphotericin B) and solution of metal salts were also screened under similar conditions for comparison. The Petri dishes were incubated for 48 h. at 37 or 28ºC, respectively. The zone of inhibition was measured in millimeters carefully. All determinations were made in duplicated manner for all compounds. The average of the two independent readings was record.

1.4. Physical Measurements

The ligand and its metal complexes were analyzed for C, H, N, Cl and M contents at the Microanalytical Laboratory, Faculty of Science, Cairo University, Egypt. Analytical and physical data of the ligand H_2L and its metal complexes are reported in Table 1. The metal ion contents of the complexes were also determined by the previously reported methods. [14-16] IR spectra of the ligand and its metal complexes were measured using KBr discs with a

Jasco FT/IR 300E Fourier transform infrared spectrophotometer covering the range 400–4000 cm^{-1} and in the 500–100 cm^{-1} region using polyethylene-sandwiched Nujol mulls on a Perkin Elmer FT-IR 1650 spectrophotometer. ^{1}H NMR spectra was obtained on 300 MHz VARIAN NMR spectrometers. Chemical shifts (ppm) are reported relative to TMS. The mass spectra were run at 70 eV with HP MODEL: MS. 5988A. The electronic spectra of the ligand and its complexes were obtained in Nujol mulls using a Shimadzu UV–240 UV–Vis recording spectrophotometer. Molar conductivities of the metal complexes in DMSO (10^{-3} M) were measured using a dip cell and a Bibby conductimeter MC1 at room temperature. The resistance measured in ohms and the molar conductivities were calculated according to the equation: $\Lambda = V \times K \times Mw/g \times \Omega$, where Λ, molar conductivity ($ohm^{-1}\ cm^{2}\ mol^{-1}$); V, volume of the complex solution (mL); K, cell constant 0.92 cm^{-1}; Mw, molecular weight of the complex; g, weight of the complex; and Ω, resistance measured in ohms. Magnetic moments at 298 K were determined using the Gouy method with $Hg[Co(SCN)_4]$ as calibrant. The solid ESR spectra of the complexes were recorded with ELEXSYS E500 Brucker spectrometer in 3-mm Pyrex Tubes at 298 K. Diphenylpicrylhydrazide (DPPH) was used as a g-marker for the calibration of the spectra.

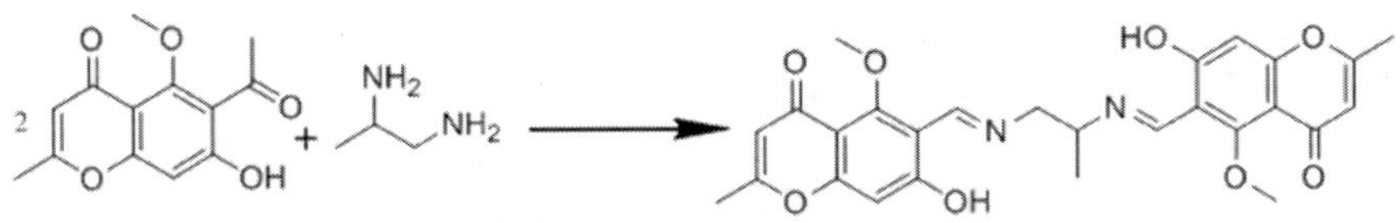

Scheme 1. Schematic representation for the formation of the ligand H_2L.

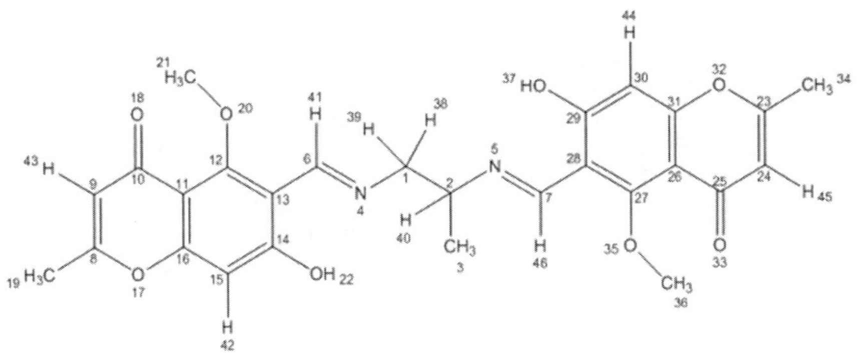

Scheme 2. Numbering scheme employed for H_2L (1).

Results and Discussion

Elemental Analysis

The elemental and physical data of the ligand H_2L and its complexes (Table 1) showed that the stoichiometry of the complexes obtained is either 1:1 or 2:1 M/L (metal: ligand).

Mass Spectra of the Ligand

The mass spectra of the Schiff base ligand H_2L revealed the molecular ion peak at m/e 506, which is coincident with the formula weight (506.5) for this ligand and supports the identity of the structure.

Infrared Spectra

The important spectral bands of the synthesized ligand as well as the complexes are presented in Tables 2. The Schiff base ligand exhibit two broad bands centered at 3250 and 3247 cm^{-1} due to ν(O–H). These bands disappeared on complexation with the metal ion in case of complexes 3, 5, 11 and 12, which indicates the deprotonation of the ligands prior to coordination through oxygen atom. [17, 18] In other complexes these bands were shifted to lower wave number (10-20 and 12-22 cm^{-1}) with decreasing their intensities indicating their coordination to the metal ion [18-20]. A strong bands around 1648 and 1651s cm^{-1} due to azomethine group of the ligand undergoes shifted to lower wave number (18-26 and 15-23 cm^{-1}) with decreasing their intensities indicating the coordination to the azomethine nitrogen to metal ion. [18, 21, 22] The band at 1218 cm^{-1} in the free ligand may be assigned to the phenolic C–O stretching. [23] In the complexes, C–O stretching vibrations are shifted to higher frequency by 7-24 cm^{-1}. This shift to higher frequency indicates the bonding of phenolic oxygen to metal ion. [24] The strong band at 1663 and 1661 cm^{-1} due to carbonyl groups of the ligand almost showed no shift in all complexes indicating that both carbonyl groups were not involved in complexation.

Some complexes included water molecules. The broad bands in the 3485-3441 cm^{-1} region are due to coordinated water or water of crystallization. The bands for water of crystallization are different from those of coordinated water. The presence of water molecules within the coordination sphere in the hydrated complexes 3, 6, 8, 10, 12 and 13 is supported by the presence of bands in the 3432-3448 , 1600-1610, 940-955 and 620-632 cm^{-1} regions due to OH stretching, HOH deformation, H_2O rocking and H_2O wagging, respectively. [25] The absence of coordinated water molecules in complexes 2, 4 and 11 was confirmed from the absence of the rocking, twisting and wagging vibrational modes which are normally activated in 970-930 cm^{-1} and 660-600 cm^{-1} regions, [26] as well as the presence of weak and broad bands in 3530-3485 cm^{-1} range indicating that the water in all of these complexes is lattice rather than coordinated [27].

Extensive IR spectral studies reported on metal aceto complexes indicate that the acetate ligand may coordinate to a metal center in either a monodentate, bidentate or bridging manner. [27] The $\nu_{asym.}(CO_2)$ and $\nu_{sym.}(CO_2)$ of the free acetate ions are at 1560 cm^{-1} and 1416 cm^{-1}, respectively. [27] In monodcntate coordination ν(C=O) is found at higher energy than $\nu_{asym.}(CO_2)$ and ν(C-O) is lower than $\nu_{sym.}(CO_2)$. As a result, the separation between the two ν(CO) bands is much larger in monodentate complexes than the free ion as found for the complex 4, 9 and 10 [22, 28] and this is supported by the appearance of two new bands in 1579-1593 and 1363-1384 cm^{-1} regions, which may be attributed to ν_{asym} (COO^-) and $\nu_{sym.}(COO^-)$, respectively. The separation value (Δ) between ν_{asym} (COO^-) and $\nu_{sym.}(COO^-)$ in these complexes was more than 200 cm^{-1} (209-216) suggesting the coordination of carboxylate group in a monodentate fashion. [28, 29] Further, the complex exhibits δ(COO) at 750 cm^{-1} which is considered diagnostic for unidentate acetates. [30] The opposite trend is observed in bidentate aceto coordination; the separation between ν(CO) is smaller than for the free ion. For bridging acetate with both oxygens coordinated as in copper (ll) acetate, however, the two ν(CO) bands are close to the free ion values. [22, 29, 31, 32].

The spectra of the complexes 2-3, 5-8 and 12-13 showed two assignable, weak intensity ν (M-CI) bands in 410-377 cm^{-1} range indicating terminal Chloro ligands [33].

Conductivity Measurements

All metal complexes are stable in air and insoluble in common organic solvents but easily soluble in DMF and DMSO. The molar conductivities of the complexes in DMF (10^{-3} M) are listed in table 1. All complexes show a non-electrolyte nature. [22, 32, 34].

Electronic Spectra and Magnetic Moments

The solid state spectrum of the ligand H_2L shows a broad band at 268 nm assignable to the phenyl ring $\pi\rightarrow\pi^*$ transitions [35, 36] and 355 nm assignable to $\pi\rightarrow\pi^*$ of C=N groups [37].

The $n\rightarrow\pi^*$ transition associated with the (C=N) moiety found around 395 nm in the spectrum of the free ligand and is generally shifted to higher energy on complexation. [38] There are also charge-transfer transitions partially responsible for the intense colors of the complexes. The bands present at energies above 333 nm in the spectra of the metal complexes are Cl→ M(II) and O→ M(II) bands. [35, 36]. The cobalt(II) complexes 2 and 6 have spectra typical of an octahedral system showing two bands, in 622-639 and 467-480 nm regions, assignable to $^4T_{1g}$ (F) → $^4A_{2g}$ (F) (ν_2) and $^4T_{1g}$ (F)→$^4T_{2g}$ (P) (ν_3) transition, respectively, indicating an octahedral structure. [39-41] The magnetic moment value (μ_{eff}) of the Co(II) complex 2 found to be 4.7 B. M., which lie in the range reported for high spin (S = 3/2) octahedral cobalt(II) complexes [42]. Complexes 3 and 12 showed two bands in 490-495 and 645-655 nm regions which are assignable to $^6A_{1g}$ → $^4A_{1g}$(G) and $^6A_{1g}$ → $^4T_{2g}$(G) transitions, respectively, suggesting its octahedral nature [41, 43].

The magnetic moment of complex 3 is 5.5 BM which is closer to the spin only value indicating an octahedral structure for the complex [44].

The nickel(ll) complex 13 have electronic spectra typical of an octahedral system showing two bands lie at 645 and 510 nm which may be assigned to $^3A_{2g}$→ $^3T_{1g}$(F) (ν_2) and $^3A_{2g}$→ $^3T_{1g}$(P) (ν_3) transitions, respectively, in an octahedral geometry. [45-47] The diamagnetic complex 11 showed shoulder at 539 nm characteristic of a square planar diamagnetic Ni(II) complexes, [48] and resemble the absorption bands observed in the spectrum of an analogous Ni(II) complexes reported by Bouwman [49].

Copper complexes 4, 8 and 10 showed signals in 567-580 and 622-635 nm regions, which were assigned to $^2B_1$→^2E and $^2B_1$→$^2B_2$ transitions indicating a distorted octahedral structure, [50] while complexes 5, 7 and 9 complexes register a broad band in 540-543 nm region due to $^2B_{1g}$ → ^2E_g transition in a square planar Cu(II) structure [51].

The magnetic moment of complex 4 is 1.82 B.M. indicating the presence of one unpaired electron. The electronic spectral results clearly indicated square planar geometry for the Cu(II) system, which should contain one unpaired electron, where μ_{eff} value would be in the range 1.8–2.1 B.M. [52].

Table 1. Analytical and physical data of the ligand H_2L and its metal complexes

No.	Ligand/ Complexes	FW	Yield (%)	Analysis (%) / Found (calcd)					Molar Conductance
				C	H	N	Cl	M	
1	H_2L $C_{27}H_{26}N_2O_8$	506.5	71	63.8(64.0)	5.4(5.2)	5.6(5.5)	-	-	-
2	$[H_2LCoCl_2].2H_2O$ $C_{27}H_{30}Cl_2CoN_2O_{10}$	672.4	65	48.0(48.2)	5.6(4.5)	4.5(4.2)	10.2(10.6)	8.7(8.8)	21
3	$LFeCl.H_2O$ $C_{27}H_{26}ClFeN_2O_9$	613.8	73	52.5(52.8)	4.4(4.3)	4.8(4.6)	5.7(5.8)	9.0(9.1)	18
4	$[H_2LCu(OC(O)CH_3)_2].3H_2O$ $C_{31}H_{38}CuN_2O_{15}$	742.2	81	50.0(50.2)	5.3(5.2)	4.1(3.8)	-	8.4(8.6)	13
5	LCu $C_{27}H_{24}CuN_2O_8$	568.0	79	57.0(57.1)	4.6(4.3)	5.0(4.9)	-	11.0(11.2)	7
6	$H_2LCo_2Cl_4(H_2O)_4$ $C_{27}H_{34}Cl_4Co_2N_2O_{12}$	838.2	74	38.6(38.7)	4.1(4.1)	3.6(3.3)	16.8(16.9)	14.0(14.1)	25
7	$H_2LCu_2Cl_4$ $C_{27}H_{26}Cl_4Cu_2N_2O_8$	775.4	84	41.5(41.8)	3.6(3.4)	3.8(3.6)	18.2(18.3)	16.2(16.4)	20
8	$H_2LCu_2Cl_4(H_2O)_4$ $C_{27}H_{34}Cl_4Cu_2N_2O_{12}$	847.5	79	38.1(38.3)	40.1(4.0)	3.6(3.3)	16.5(16.7)	14.9(15.0)	22
9	$H_2LCu_2(OC(O)CH_3)_4$ $C_{35}H_{38}Cu_2N_2O_{16}$	869.8	85	48.0(48.3)	4.5(4.4)	3.4(3.2)	-	14.5(14.6)	16
10	$H_2LCu_2(OC(O)CH_3)_4(H_2O)_4$ $C_{35}H_{46}Cu_2N_2O_{20}$	941.8	84	44.5(44.6)	5.1(4.9)	3.1(3.0)	-	13.2(13.5)	14
11	$[LNi].H_2O$ $C_{27}H_{26}N_2NiO_9$	581.2	88	55.6(55.8)	4.7(4.5)	5.0(4.8)	-	9.9(10.1)	8
12	$LFe_2Cl_4(H_2O)_4$ $C_{27}H_{32}Cl_4Fe_2N_2O_{12}$	830.1	69	39.0(39.1)	4.0(3.9)	3.6(3.4)	16.9(17.1)	13.4(13.5)	27
13	$H_2LNi_2Cl_4(H_2O)_4$ $C_{27}H_{34}Cl_4N_2Ni_2O_{12}$	837.8	73	38.5(38.7)	4.3(4.1)	3.4(3.3)	16.8(16.9)	13.8(14.0)	22

Molar Conductance $=\Lambda_m$ $(\Omega^{-1}cm^2mol^{-1})$.

Table 2. IR frequencies of the bands (cm-1) of ligand H_2L and its complexes and their assignments

No.	Ligand/ Complexes	ν(H_2O)	ν(OH)	ν(C=O)	ν(C=N)	ν(C–O)	ν_s(Coo), ν_{as}(Coo), (Δ)	ν(M-N)	ν(M-O)	ν(M-Cl)
1	H_2L	-	3250m, 3247m	1663s, 1661s	1648s, 1651s	1218m	-	-	-	-
2	$[H_2LCoCl_2].2H_2O$	3465b	3230m 3225sh	1662m	1631m	1225m	-	450m	410m	377
3	$LFeCl.H_2O$	3432b	-	1661m	1633m	1239m	-	495m	390m	398
4	$[H_2LCu(OC(O)CH_3)_2].3H_2O$	3485b	3236m 3225m	1663m	1628m	1228m	1579, 1363, 216	440w	400w	-
5	LCu	-	-	1660m	1625s	1241m	-	480m	380w	-
6	$H_2LCo_2Cl_4(H_2O)_4$	3445b	3235m, 3228m	1664m	1630m	1223m	-	460w	355w	405
7	$H_2LCu_2Cl_4$	-	3234m, 3230sh	1663m	1629m	1227m	-	450w	350w	375
8	$H_2LCu_2Cl_4(H_2O)_4$	3438b	3234m 3228m	1662m	1633m, 1627m	1224m	-	434m	364m	410
9	$H_2LCu_2(OC(O)CH_3)_4$	-	3240m 3230m	1660m	1628m	1229m	1586, 1376, 210	448m	394m	-
10	$H_2LCu_2(OC(O)CH_3)_4(H_2O)_4$	3448b	3240m 3235m	1663m	1630	1228m	1593, 1384, 209	445m	368m	-
11	$[LNi].H_2O$	3530b	-	1662b	1631m 1628sh	1242m	-	475m	355m	-
No.	Ligand/ Complexes	ν(H_2O)	ν(OH)	ν(C=O)	ν(C=N)	ν(C–O)	ν_s(Coo), ν_{as}(Coo), (Δ)	ν(M-N)	ν(M-O)	ν(M-Cl)
12	$LFe_2Cl_4(H_2O)_4$	-	3238m 3228m	1660b	1629m	1225m	-	441w	387m	400
13	$H_2LNi_2Cl_4(H_2O)_4$	3444b	3239m 3227m	1663b	1630	1227m	-	437m	380m	405

m=medium, s=strong, b=broad, w=week, sh=shoulder.

Table 3. The electronic absorption spectral bands (nm) and magnetic moment (B.M) for the ligand H_2L and its Metal complexes

No.	Ligand/ Complexes	π–π*, n–π*, and CT bands	d–d bands	μ_{eff} in BM
1	H_2L	268, 325, 395	-	-
2	$[H_2LCoCl_2].2H_2O$	278, 345, 390, 410	480, 639	4.7
3	$LFeCl.H_2O$	260, 366, 387, 407	490, 655	5.5
4	$[H_2LCu(OC(O)CH_3)_2].3H_2O$	268, 355, 392, 425	580, 635	1.71
5	LCu	255, 340, 380, 405	540	1.82
6	$H_2LCo_2Cl_4(H_2O)_4$	310, 335, 385, 410	467, 622	3.4
7	$H_2LCu_2Cl_4$	261, 342, 397, 400	541	1.53
8	$H_2LCu_2Cl_4(H_2O)_4$	299, 354, 396, 408	576, 630	1.49
9	$H_2LCu_2(OC(O)CH_3)_4$	288, 349, 397, 412	543	1.48
10	$H_2LCu_2(OC(O)CH_3)_4(H_2O)_4$	259, 355, 380, 422	574, 530	1.56
11	$[LNi].H_2O$	266, 345, 392, 402	539sh	-
12	$LFe_2Cl_4(H_2O)_4$	258, 364, 390, 413	495, 645	3.2
13	$H_2LNi_2Cl_4(H_2O)_4$	267, 355, 381, 400	510, 645	1.9

The magnetic moment of complexes 6-10 and 12-13 were smaller than the calculated value for two metal ions and this may be due to antiferromagnetism between the two ion-centers. [22, 32, 53] The structure representation of metal complexes is present in figure 1.

ESR Spectra

The ESR spectra of polycrystalline samples of copper(II) complexes (4-5 and 7-10) at room temperatures (298 K) were of axial shape with $g_{//} > g_{\perp}$ characteristic of complexes with $^{2}B_{1(dx2-y2)}$ orbital ground state. [54] The average g values were calculated according to the equation: $g_{av} = 1/3[g_{//} + 2 g_{\perp}]$, Table 4. The complexes show $g_{\|}>g_{\perp}>g_e$, indicating a d_{x2-y2} ground state, and spectral features were characteristic of axial symmetry. [55] All complexes exhibit $g_{//} < 2.3$, suggesting covalent characters around the copper in present complexes. Kivelson and Neiman [56] have reported that a $g_{//}$ value greater than 2.3 indicates ionic character. The g-values were related by the expression, [22, 53] $G = (g_{\|}-2)/(g_{\perp}-2)$, if $G > 4.0$ then local tetragonal axes were aligned parallel or only a slightly misaligned, if $G < 4.0$, significant exchange coupling is present, table 4. Only complexes 4 and 5 showed $G > 4.0$ indicate that there is no direct copper-copper interaction in the solid state of these complexes. In case of other copper complexes, the G value is <4, which suggests considerable exchange interaction in these solid complexes. These complexes showed a half field signal, observed at 1600 G due to the $\Delta m_s = +2$ transitions, proving the Cu-Cu interaction but the absence of such signal in case of complexes 4 and 5 rules out any Cu-Cu interaction in these tow complexes. [57] The $g_{//}/A_{//}$ is taken as an indication for the stereochemistry of the copper(II) complexes. Addison [58] has suggested that this ratio may be an empirical indication of the tetrahedral distortion of the square planar geometry. The values of $g_{//}/A_{//}$ quotient in the range 105–135 cm^{-1} are expected for copper complexes within perfectly square based geometry and those higher than 150 cm^{-1} for tetrahedrally distorted complexes. The values of $g_{//}/A_{//}$ for complexes 5, 7 and 9 lie in the range 126-130 indicating a square planar geometry while the value for complexes 4, 8 and 10 lie in the range 195-214 indicating an octahedral geometry around copper ion.

Table 4. ESR parameters for copper(II) complexes

No.	Complexes	$g_{//}$	$g_{\perp}$	$g_{//}/A_{//}$	G	$g_{av.}$
4	$[H_2LCu(OC(O)CH_3)_2].3H_2O$	2.22	2.040	195	5.5	2.10
5	LCu	2.15	2.030	130	5.0	2.07
7	$H_2LCu_2Cl_4$	2.10	2.037	127	2.7	2.10
8	$H_2LCu_2Cl_4(H_2O)_4$	2.15	2.040	214	3.8	2.13
9	$H_2LCu_2(OC(O)CH_3)_4$	2.09	2.029	126	3.1	2.09
10	$H_2LCu_2(OC(O)CH_3)_4(H_2O)_4$	2.14	2.039	204	3.6	2.10

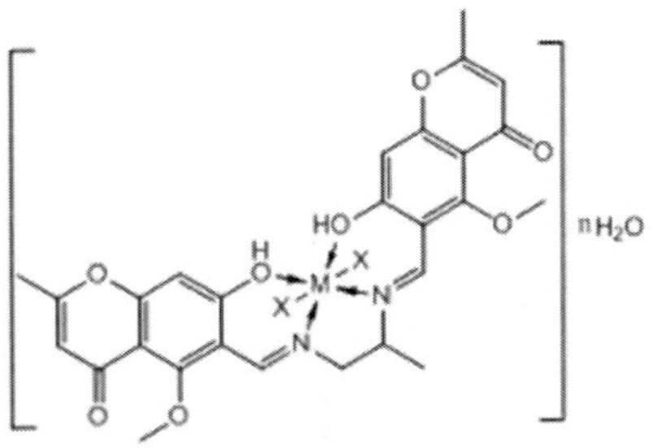

Complex **2**: $[H_2LCoCl_2].2H_2O$, M=Co, X=Cl, n=2
Complex **4**: $[H_2LCu(OC(O)CH_3)_2].3H_2O$, M=Cu, X=$OC(O)CH_3$, n=3

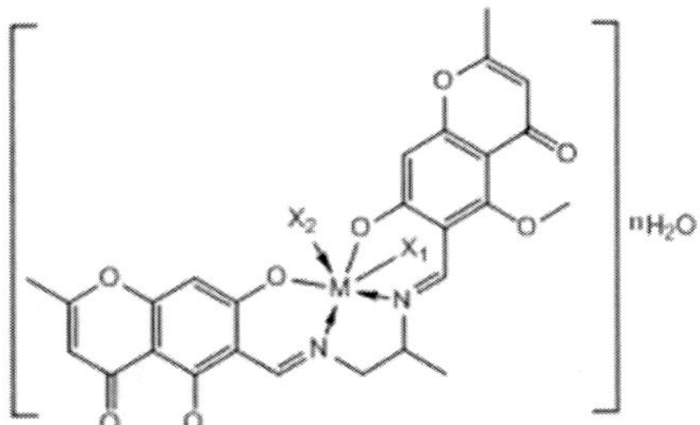

Complex **3**: $LFeCl.H_2O$, M=Fe, X_1=Cl, X_2=H_2O, n=none
Complex **5**: LCu, M=Cu, X_1=X_2=None, n=none
Complex **11**: $[LNi].H_2O$, M=Ni, X_1=X_2=none, n=1

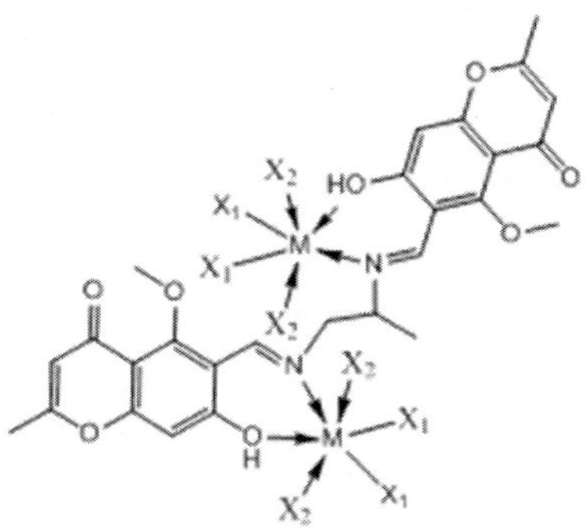

Complex **6**: $H_2LCo_2Cl_4(H_2O)_4$, M=Co, X_1=Cl, X_2=H_2O
Complex **7**: $H_2LCu_2Cl_4$, M=Cu, X_1=Cl, X_2=none
Complex **8**: $H_2LCu_2Cl_4(H_2O)_4$, M=Cu, X_1=Cl, X_2=H_2O
Complex **9**: $H_2LCu_2(OC(O)CH_3)_4$, M=Cu, X_1=$OC(O)CH_3$, X_2=none

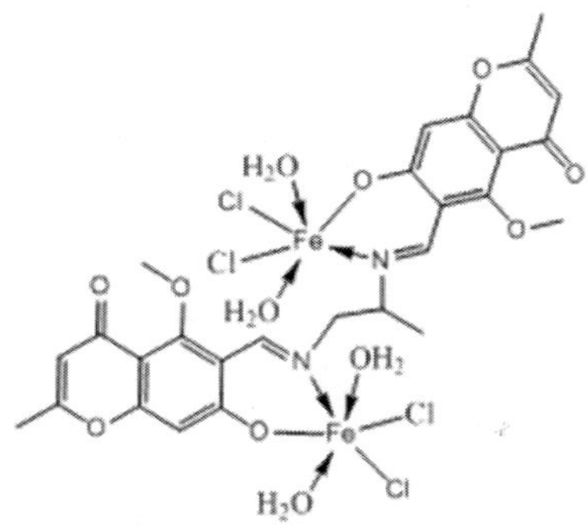

Complex **10**: $H_2LCu_2(OC(O)CH_3)_4(H_2O)_4$, M=Cu, X_1=$(OC(O)CH_3)$, X_2=H_2O
Complex **12**: $LFe_2Cl_4(H_2O)_4$
Complex **13**: $H_2LNi_2Cl_4(H_2O)_4$, M=Ni, X_1=Cl, X_2=H_2O

Figure 1. The proposed structures of metal complexes of the ligand H_2L.

Biological Activity

The results of biological activity tested for ligands and their complexes are given in figures 2, 3. Diameter of inhibition zone (mm) including the disc diameter was measured for each treatment. The ligands showed no activity at all.

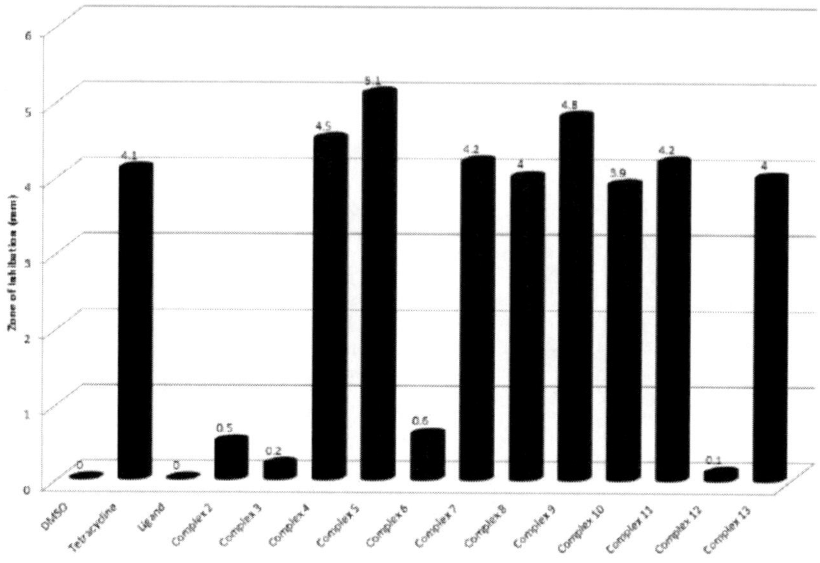

Figure 2. Antibacterial activity of the ligand and its metal complexes against gram-negative bacterium (E. coli).

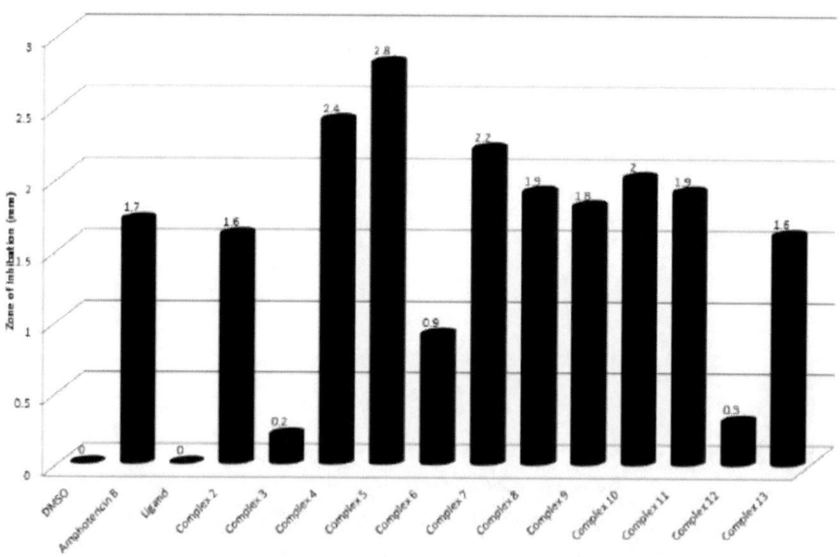

Figure 3. Antifungal activity of the ligand and its metal complexes against Fungus (Aspergillus niger).

The Cu and Ni complexes exhibited the maximum antibacterial and antifungal activities, while Co and Fe complexes showed moderate antibacterial and antifungal activities in this study. The results indicate that the complexes show more activity and the ligands do not have any activity against some organisms under identical experimental conditions. This would suggest that the chelation could facilitate the ability of a complex to cross a cell membrane and can be explained by Tweedy's chelation theory. [59] Chelation considerably reduces the polarity of the metal ion mainly because of partial sharing of its positive charge with the donor groups and possible electron delocalization over the whole chelate ring. Such chelation could also enhance the lipophilic character of the central metal atom, which subsequently favours its permeation through the lipid layer of the cell membrane [60].

CONCLUSION

The ligand H_2L and its corresponding Fe(III), Co(II), Cu(II) and Ni(II) complexes have been synthesized and spectrally characterized. The ligand acted either as bibasic tetradentate, neutral tetradentate. All prepared complexes showed either octahedral or square planar geometries. Complexes 6-10 and 12-13 showed magnetic moments lower than the spin-only value indicating spin-exchange interactions between metal ions. This was more confirmed in case of copper complexes 7-10, which showed G value < 4 suggesting considerable exchange interaction between copper centers in this solid complexes. The values of $g_{//}/A_{//}$ for complexes 5, 7 and 9 lie in the range 126-130 indicating a square planar geometry while the value for complexes 4, 8 and 10 lie in the range 195-214 indicating a octahedral geometry around copper ion. The antibacterial and antifungal activities of the compounds showed that, some of metal complexes exhibited a greater inhibitory effect than the ligand, corresponding aqueous ions, and standard drugs as tetracycline (bacteria) and amphotericin B (fungi).

REFERENCES

[1] Knof U.; von Zelewsky A., "Predetermined Chirality at Metal Centers" *Angew. Chem. Int. Ed.*, 38(1999) 302-322.

[2] Knight P.D.; Scott P., "Predetermination of chirality at octahedral centres with tetradentate ligands: prospects for enantioselective catalysis" *Coord. Chem. Rev.*, 242 (2003) 125-143.

[3] You Z-L; Zhu H-L; Liu W-S, "Solvolthermal Syntheses and Crystal Structures of Three Linear Trinuclear Schiff Base Complexes of Zinc(II) and Cadmium(II) Z." *Anorg. Allg. Chem.* 630(2004) 1617- 1622.

[4] Golcu, A.; Tumer, M.; Demirelli, H.; Wheatley, R.A., "Cd(II) and Cu(II) complexes of polydentate Schiff base ligands: synthesis, characterization, properties and biological activity" *Inorganica Chimica Acta*, 358(6), (2005). 1785-1797.

[5] Ziessel R., "Schiff-based bipyridine ligands. Unusual coordination features and mesomorphic behaviour" *Coord. Chem. Rev.*, 216 (2001) 195-223.

[6] Parashar R.K.; Sharma R.C.; Kumar A.; Mohan G., "Stability studies in relation to IR data of some schiff base complexes of transition metals and their biological and pharmacological studies" *Inorg. Chim. Acta*, 151(1988) 201-208.

[7] J.R. Lancaster (Ed.), *The Bioinorganic Chemistry of Nickel,* VCH, New York, 1988.

[8] Chandra S.; Sangeetika X., "EPR, magnetic and spectral studies of copper(II) and nickel(II) complexes of schiff base macrocyclic ligand derived from thiosemicarbazide and glyoxal" *Spectrochim. Acta A,* 60 (2004)147-153.

[9] Opstal T.; Verpoort F, "Synthesis of Highly Active Ruthenium Indenylidene Complexes for Atom-Transfer Radical Polymerization and Ring-Opening-Metathesis Polymerization" *Angew. Chem. Int. Ed.*, 42 (2003) 2876-2879.

[10] De Clercq B.; Lefebvre F.; Verpoort F., "Immobilization of multifunctional Schiff base containing ruthenium complexes on MCM-41" *Appl. Catal. A* 247 (2003) 345-364.

[11] Lambert S.L.; Spiro C.L.; Gagne R.R.; Hendriekson D.N. "Binuclear complexes of macrocyclic ligands. Variation of magnetic exchange interaction in a series of heterobinuclear CuII-MII complexes" *Inorg. Chem.*, 21 (1982) 68-72.

[12] Hodnett E.M.; Dunn W.J., "Improved synthesis of the antibiotic, hexahydrospinamycin" *J. Med. Chem.,* 15 (1972) 339-340.

[13] J.G. Collee, J.P. Duguid, A.G. Farser, B.D. Marmion (editors) "Practical Medical Microbiology" New York, Churchill Livingstone (1989).

[14] F.J. Welcher, *The Analytical Use of EDTA*, Van Norstrand, USA, 1958.

[15] A.I. Vogel, *A Text Book of Quantitative Inorganic Analysis,* fourth ed., Longman, London, 1978.

[16] Z. Holzbecher, L. Divis, M. Kral, L. Sucha, F. Vlacil, *Handbook of Organic Reagents in Inorganic Analysis*, John Wiley, New York, 1976.

[17] Kannan S.; Ramesh R., "Synthesis, characterization, catalytic oxidation and biological activity of ruthenium(III) Schiff base complexes derived from 3-acetyl-6-methyl-2H-pyran-2,4(3H)-dione" *Polyhedron,* 25 (2006) 3095-3103.

[18] Youssef N.S.; El-Zahany E.; Barsoom B.N.; El-Seidy A.M.A., "Synthesis and characterization of copper(II), cobalt(II), nickel(II), and iron(III) complexes with two diamine Schiff bases and catalytic reactivity of a chiral diamine cobalt(II) complex" *Transition Met. Chem.,* 34 (2009) 905- 914.

[19] K. Nakamoto "Infrared Spectra of Inorganic and Coordination Compounds" 2nd Edit., Wiley - Interscience. New York, 1970.

[20] Abd El-Wahab Z.H.; Mashaly M.M.; Salman A.A.; El-Shetary B.A.; Faheim A.A., "Co(II), Ce(III) and UO2(VI) bis-salicylatothiosemicarbazide complexes: Binary and ternary complexes, thermal studies and antimicrobial activity" Spectrochim. *Acta Part A,* 60 (2004) 2861-2873.

[21] Karvembu R.; Jayabalakrishnan C.; Dharmaraj N.; Renukadevi S.V.; Natarajan K. "Binuclear ruthenium(III) complexes: synthesis, characterisation, catalytic activity in aryl–aryl couplings and biological activity" *Trans. Met. Chem.,* 27 (2002) 631-638.

[22] Youssef N.S.; El-Zahany E.; El-Seidy A.M.A.; Caselli A.; Cenini S., "Synthesis and characterization of some transition metal complexes with a novel Schiff base ligand and their use as catalysts for olefin cyclopropanation" *J. Mol. Catal. A: Chem.* 308(1, 2) (2009) 159-168.

[23] Keskioglu E.; Gunduzalp A.B.; Cete S.; Hamurcu F.; Erk B., "Cr(III), Fe(III) and Co(III) complexes of tetradentate (ONNO) Schiff base ligands: Synthesis, characterization, properties and biological activity" *Spectrochim Acta A*, 70 (2008) 634-640.

[24] Ramesh R, Maheswaran S "Synthesis, spectra, dioxygen affinity and antifungal activity of Ru(III) Schiff base complexes" (2003) *J. Inorg. Biochem.* 96:457-462.

[25] Teotia M.; Gurthu J.N.; Rama V.B., "Dimeric 5-and 6-coordinate complexes of tri and tetradentate ligands" *J. Inorg. Nucl. Chem.*, 42 (1980) 821-831.

[26] El-Dissouky A.; Fahmy A.; Amer A., "Complexing ability of some γ-lactone derivatives. Thermal, magnetic and spectral studies on cobalt(II), nickel(II) and copper(II) complexes and their base adducts" *Inorg. Chim. Acta*, 133(1987) 311-316.

[27] K. Nakamoto, "Infrared Spectra of Inorganic and coordination compounds", John Wiley and Sons, New York, P 173 (1965).

[28] Boghaei D.M.; Gharagozlou M., "Spectral characterization of novel ternary zinc(II) complexes containing 1,10-phenanthroline and Schiff bases derived from amino acids and salicylaldehyde-5-sulfonates" *Spectrochim. Acta A* 67(2007) 944-949.

[29] Shauib N.M.; Elassar A-ZA.; El-Dissouky A., "Synthesis and spectroscopic characterization of copper(II) complexes with the polydentate chelating ligand 4,4′-[1,4-phenylenedi(nitrilo)dipente-2-one" *Spectrochim. Acta Part A*, 63 (2006) 714-722.

[30] K. Nakamota, "Infrared and Raman Spectra of Inorganic and coordination compounds", 4th Edn, John Wiley and Sons, New York, 1986.

[31] El-Shazly R.M.; Al-Hazmi G.A.A.; Ghazy S.E.; El-Shahawi M.S.; El-Asmy A.A., "Spectroscopic, thermal and electrochemical studies on some nickel(II) thiosemicarbazone complexes" *Spectrochim. Acta A* 61 (2005) 243-252.

[32] Youssef N.S.; El-Zahany E.; El-Seidy A.M.A.; Caselli A.; Fantauzzi S.; Cenini S., "Synthesis and characterisation of new Schiff base metal complexes and their use as catalysts for olefin cyclopropanation" *Inorg. Chim. Acta*, 362 (2009) 2006-2014.

[33] Ferrari M.B.; Fara G.G.; Lafranchi M.; Pelizzi C.; Tarasconi M., "Synthesis, spectroscopic and structural characterization of chlorobis(methyl pyruvate thiosemicarbazone)copper(I) and chlorobis(triphenylphosphine) (methyl pyruvate thiosemicarbazone) copper(I) toluene solvate (2/1)" *Inorg. Chem. Acta*, 181 (1991) 253-262.

[34] Geaey W.J.; "The use of conductivity measurements in organic solvents for the characterisation of coordination compounds" *Coord. Chem. Rev.*, 7 (1971) 81-122.

[35] Khalil M.M.H.; Aboaly M.M.; Ramadan R.M., "Spectroscopic and electrochemical studies of ruthenium and osmium complexes of salicylideneimine-2-thiophenol Schiff base" *Spectrochim. Acta Part A*, 61 (2005) 157-161.

[36] Zhang H-Y.; Wu K-Z.; Zhang J-J.; Xu S-L.; Ren N.; Bai J-H.; Tian L, "Synthesis, crystal structure and thermal decomposition kinetics of the complex [Sm(BA)3bipy]2" *Synthetic Metals*, 158 (2008) 157-164.

[37] Wu G.; Wang X.; Li J.; Zhao N.; Wei W.; Sun Y, "A new route to synthesis of sulphonato-salen-chromium(III) hydrotalcites: Highly selective catalysts for oxidation of benzyl alcohol to benzaldehyde" *Catal. Today*, 131 (2008) 402-407.

[38] Mostafa S.; Ikeda S.; Ohtani B, "Transition metal Schiff-base complexes chemically anchored on Y-zeolite: their preparation and catalytic epoxidation of 1-octene in the suspension and phase boundary systems" *J. Mol. Catal. A: Chem.*, 225 (2005) 181-188.

[39] Mohamed G.G.; El-Wahab Z.H.A., "Mixed ligand complexes of bis(phenylimine) Schiff base ligands incorporating pyridinium moiety: Synthesis, characterization and antibacterial activity" *Spectrochim. Acta Part A*, 61(6) (2005) 1059-1068.

[40] Tascioglu S, Yalcin B, Nasrullayeva TM, Andac O, Buyukgungor O, Aydin A, Medjidov AA (2006) *Polyhedron*, 25(6):1279.

[41] B. P. Lever, *"Inorganic Electronic Spectroscopy"*, Elsevier, Amsterdam, 341, (1968).

[42] Shakir M.; Azim Y.; Chishti H-T-N.; Parveen S., "Synthesis, characterization of complexes of Co(II), Ni(II), Cu(II) and Zn(II) with 12-membered Schiff base tetraazamacrocyclic ligand and the study of their antimicrobial and reducing power" (2006) *Spectrochim. Acta Part A*, 65:490-496.

[43] Kumar K.G.; John K.S., "Complexation and ion removal studies of a polystyrene anchored Schiff base" *Reactive and Functional Polymers 66*(12) (2006) 1427-1433.

[44] F.A. Cotton, G. Wilkinson, Advanced Inorganic Chemistry, John Wiley and Sons, New York, 1998.

[45] Mostafa SI, Ikeda S, Ohtani B "Transition metal Schiff-base complexes chemically anchored on Y-zeolite: their preparation and catalytic epoxidation of 1-octene in the suspension and phase boundary systems" *J. Mol. Catal. A,* 225 (2005)181-188.

[46] Mohamed G.G.; Omar M.M.; Hindy A.M.M., "Synthesis, characterization and biological activity of some transition metals with Schiff base derived from 2-thiophene carboxaldehyde and aminobenzoic acid" *Spectrochim. Acta Part A*, 62 (2005) 1140-1150.

[47] Mohamed G.G.; El-Wahab Z.H.A., "Mixed ligand complexes of bis(phenylimine) Schiff base ligands incorporating pyridinium moiety:

Synthesis, characterization and antibacterial activity" *Spectrochim. Acta Part A,* 61 (2005) 1059-1068.

[48] Amirnasr M, Schenk KJ, Meghdadi S, Morshedi M "Synthesis, characterization and X-ray crystal structures of [Ni(Me-sal)2dpt] and [Ni(Me-sal)dpt]Cl" *Polyhedron* 25 (2006) 671-677.

[49] Bouwman E.; Henderson R.K.; Reedijk J.; Veldman N.; Spek A.L., "Synthesis of the ligand N,N′-bis(2-tert-butylthiobenzenylidene)-diethylenetriamine; its reactivity with nickel(II) salts" *Inorg. Chim. Acta* 287 (1999) 105-108.

[50] B. P. Lever *"Inorganic Electronic Spectroscopy"* 2nd edn, Elsevier, Amsterdam, 1984.

[51] Kumar K.G.; John K.S., "Complexation and ion removal studies of a polystyrene anchored Schiff base" *React. Funct. Polym.* 66 (2006) 1427-1433.

[52] Nair M.S.; Joseyphus R.S., "Transition metal Schiff-base complexes chemically anchored on Y-zeolite: their preparation and catalytic epoxidation of 1-octene in the suspension and phase boundary systems" *Spectrochim Acta* A 70 (2008) 749-188.

[53] Youssef N.S.; El-Zahany E.; El-Seidy A.M.A., "Synthesis and characterization of new schiff base metal complexes and their use as catalysts for olefin cyclopropanation" Phosphorus, Sulfur Silicon *Relat. Elem*. 185 (2010) 785-798.

[54] Chandra S.; Kumar U., "Spectral and magnetic studies on manganese(II), cobalt(II) and nickel(II) complexes with Schiff bases" *Spectrochim. Acta* A, 61 (2005) 219.

[55] Patel R.N.; Singh N.; Gundla V.L.N., "Synthesis, structure and properties of ternary copper(II) complexes of ONO donor Schiff base, imidazole, 2,2′-bipyridine and 1,10-phenanthroline" *Polyhedron,* 25 (2006) 3312-3318.

[56] Kivelson .D; Neiman R., "ESR Studies on the Bonding in Copper Complexes" *J. Chem. Phys.* 35 (1961) 149-156.

[57] Cozar, O.; David, L.; Chiş, V.; Cosma, C.; Znamirovschi, V.; Damian, G.; Bratu I.; Bora Gh., "ESR study of some solvent effects of the Cu(II)-aspirinate complex" *Appl. Magn. Reson*., 8 (1995) 235-242.

[58] A.W. Addison, in: K.D. Karlin, J. Zubieta (Eds.), *Copper Coordination Chemistry Biochemical and Inorganic Perspectives,* Adenine Press, New York, 1983.

[59] Fahmi N.; Gupta I.J.; Singh R.V., "Sulfur bonded palladium(ii) and platinum(ii) complexes of biologically potent thioamides" *Phosphours Sulfur and Silicon*., 132 (1998) 1-8.

[60] Tumer M.; Ekinci D.; Tumer F.; Bulut A., "Synthesis, characterization and properties of some divalent metal(II) complexes: Their electrochemical, catalytic, thermal and antimicrobial activity studies" *Spectrochimica Acta* A 67(3-4) (2007) 916.

In: Smart Nanoobjects
Editor: Kirill Levine
ISBN: 978-1-62808-741-3

Chapter 2

SYNTHESIS OF 2-AMINOTHIAZOLE DYE AND ITS METAL COMPLEXES AS AGENTS FOR ORGANIC/INORGANIC HETEROJUNCTION

Fathy A. El-Saied[1*], Ehab Abdel-Latif[1], Ahmed M. Nawar[2], Ahmed S. Radwan[1] and Atef A. Hamed[1]

[1]Chemistry Department, College of Science, Qassim University, KSA, Saudi Arabia,
[2]Thin Film Laboratory, Physics Department, Faculty of Science, Suez Canal University, Ismailia, Egypt

ABSTRACT

2-Acetylamino-5-phenylazo-thiazole (HL) was synthesized by acetylation of 2-amino-5-phenylazo-thiazole with acetic anhydride. Metal complexes of Cu(II), Co(II), Ni(II) and Fe(III) with the ligand have been prepared and characterized using variety of analytical, spectral, magnetic and thermal techniques. The obtained data show that the reaction of ligand with metal ions led to the formation of 1:1 or 1:2 (metal: ligand) metal complexes were formed. The results also show that in Cu(II)

* Corresponding Author: Fathy A. El-Saied, E-mail address: fathi_elsaied@yahoo.com.

complex, the ligand behaves as a monobasic tridentate ligand coordinating via the azo nitrogen atom, sulfur atom and the enolic oxygen atom (NSO mode).Whereas, in case of complexes of Co(II), Ni(II) and Fe(III), the ligand behaves as a neutral bidentate, coordinating through the azo nitrogen and sulfur atoms (NS mode). Hybrid heterojunction cells based on the spin-coated dye and its metal complexes as the organic semiconductor on p-Si wafers as the inorganic semiconductor have been fabricated and investigated at room temperature 293 °K. The AFM atomic force microscopy measurements showed that the spin-coated dye and its metal complexes contain a wide range of nano-sized particles. The fabricated devices showed rectification behavior like diode. The diode parameters have been studied using current–voltage (I–V) characteristics in dark and under illumination at room temperature 293 °K. The photovoltaic parameters of all prepared Au/ spin-coated ligand or complex /p-Si/Al devices were investigated under illumination with a light intensity of 6 mW/cm^2. One of the synthesized materials showed a promising behavior for the photovoltaic applications with efficiency 3.3%.

Keywords: Thiazole; Diazonium chloride; Dyes; metal complexes; Organic/inorganic heterojunction; Photovoltaic

1. Introduction

Metal–semiconductor junctions have been commonly used in microelectronic devices and in creating important circuit elements including ohmic contacts, Schottky diodes, and metal semiconductor field-effect transistors. Upon the introduction of organic modification to semiconductor materials, research has been very active with an ultimate goal to molecularly manipulate the formation, operation, and tenability of metal–molecule–semiconductor devices [1]. Control over the electronic properties of semiconductors and a metal is a central issue for their use in electronic devices [2]. The use of molecules to modify and tailor material properties is attractive in electronic devices because of the molecules' functional variety and flexibility. The use of molecules to control electron transport is an interesting possibility, not least because of the anticipated role of molecules in future electronic devices [3]. It has been demonstrated that molecules can control the electrical characteristics of conventional metal–semiconductor junctions [2, 4], apparently without the need for electrons to be transferred onto and through the molecules. Diodes have been modified by adsorbing small molecules onto single crystals of n-type GaAs semiconductor [4]. Organic materials have been also intensively investigated due to their wide range of applications in electronics technology [5-9] disordered Semiconductors. Organic semiconductors are being explored for a number of exciting applications, such as organic light-emitting diodes (OLEDs), organic thin film transistors (OTFTs), and organic photovoltaics (OPVs). A salient feature distinguishing organic (including polymeric) semiconductors from their inorganic counterparts is the strong tendency of charge carrier localization in the former. The effects of carrier localization are reflected by the dominant role of polarons due to electronic nuclear coupling. The inherent tendency for carrier localization is enhanced by the presence of energetic disorder, which is the result of structural inhomogeneities and chemical impurities, leading to charge carrier trapping in some cases. Thus, we can view organic semiconductors as members of the disordered semiconductor family [10]. Interfacial layers are used between the photoactive BHJ layer and the electrodes in order to adjust the energy levels between the electrode and the active layer and to enhance charge derivatives show biological significance, e.g., they are found in the penicillin and vitamin B1 molecules and in coenzyme cocarboxylase [11]. The chemistry extraction [12] and planarise the surface of the underlying electrode [13]. Thiazole and its of 2-thiazolines, including new methodologies for their preparation, and recent applications, such as their growing use in organic synthesis in the biological field and asymmetric catalysis as ligands has been recently reviewed [14]. 2-Aminothiazoles are known mainly as biologically active compounds with a broad range of activity and as intermediates in the synthesis of antibiotics and dyes [15]. Several papers have been published on the use of these compounds as antimicrobial [16], antifungal [17], anti-inflammatory activity [18], anesthetic [19] and antiviral drugs [20]. 2-Aminothiazoles and its derivatives are also used in the syntheses of various types of dyes [21-24]. Furthermore, high-density optical data storage has been a subject of extensive research in the past decade, In general, cyanine dyes, phthalocyanine dyes, and metal-azo complex dyes are used in the recording layer of DVD-R (Digital Versatile Disc –Recordable) discs. It has been reported that the new technology, which employs 405 nm blue-violet diode lasers, requires a new optical recording medium matching 405 nm wavelength laser. In comparison with the dyes themselves, metal azo dyes are more light stable, allowing easier control to the wavelength by selection of the appropriate azo compounds and the ease with which the absorption band may be tuned by varying the substituents, one of the many applications of azo compounds is the optical data storage. In other words, the thermal substituent groups, and have good thermal stability [25-29] Because of the good thermal properties and suitable absorption band of azo compounds are essential features in relation to their application at high-density optical recording materials. In this work, 2-acetylamino-5-phenylazo-thiazole (HL) ligand and its Cu(II), Co(II), Ni(II) and Fe(III) complexes were synthesized and characterized by using analytical , spectral, magnetic and thermal measurements. a hybrid solar cell was designed as shown schematically in Figure 1a. The electrical properties of heterojunctions were characterized by measuring the current density–voltage (J–V) curve in the dark and under illumination. The J–V curve in the dark is the easy way to estimate the quality of the junction, grid and contact resistances. Also, the basic parameters including open-circuit voltage (V_{oc}), short-circuit current density (J_{sc}), fill factor (FF) are usually determined through three points on the J–V curve under illumination. Analysis of the J–V curve gives not only the basic parameters, but also the extracted parameters such as reverse saturation current, J_o, series resistance, R_s, shunt resistance, R_{sh}, and ideality factor, n. These parameters can be applied in turn to improve the device quality during fabrication and the solar cell performance [30-33].

2. Experimental

2.1. Instrumentation

All melting points were measured on an electrothermal Gallenkamp melting apparatus. Elemental analyses were carried out at the Microanalytical Unit, Faculty of Science, University of Mansoura, Egypt; the results were in satisfactory agreement with the calculated values. UV–visible spectra were recorded with a Perkin-Elmer Lambda 551 S spectrometer in the solid state dispersed on njull mull. IR spectra (KBr discs) were determined on a Mattson 5000 FTIR spectrometer). The 1H NMR spectra were acquired using a Bruker WP 300 spectrometer at 300 MHz using TMS as an internal standard. Thermogravimetric analyses (TGA) of the solid complexes was recorded using Shimadzu TG-50 Thermogravimetric Analyzer with heating rate 10°C / min under nitrogen atmosphere, in the range 25-1000°C. Magnetic susceptibilities of the complexes were measured at room temperature in the polycrystalline state in a borosilicate tube using Johnson Matthey Magnetic Susceptibility Balance. Magnetic susceptibility was measured and calculated by using the following equation [34].

$$\mu_{eff} = 2.828\ (\chi_n \times T)^{1/2}$$
$$\chi_a = [1.115\ L(R\text{-}R_o)]/W\times 109\ ,\ \chi_m = \chi_a \times MW\ ,\ \chi_n = \chi_m - D$$

where: χ_a = mass susceptibility, L = sample length in cm, R = tube + sample reading ,R_o = empty tub reading ,W = mass of the sample ,χ_m = molar susceptibility, MW = Molecular weight , χ_n = corrected molar susceptibility D = diamagnetic correction, μ_{eff} = effective magnetic moment, T = room temperature in Kelvin (°k)

The theoretical effective magnetic moment value calculated using the equation

$$\mu_{eff} = [n(n+2)]^{1/2},$$

where: μ_{eff} = theoretical effective magnetic moment; n = the number of the unpaired electrons. Diamagnetic corrections were made using Pascal's constants [35]. Molar conductance measured by using Hanna HI8733 conductivity meter 1999.9 µs temperature coefficient = 1.

2.2. Synthesis of the Organic Dye (HL)

2.2.1. Synthesis of 2-Amino-5-Phenylazo-Thiazole (2)

A solution of phenyl diazonium chloride was prepared by adding cold sodium nitrite solution (0.7 g in 10 ml H_2O) to a cold suspension of 0.93 ml aniline in 3 ml concentrated HCl with stirring.

The freshly prepared solution was added with continuous stirring to a cold solution (0-5°C) of 2-aminothiazole (0.01 mol) in 30 ml ethanol and 4.0 g sodium acetate. The reaction mixture was stirred at 0-5°C for 2 hours, diluted with water, and then filtered. 2-Amino-5-phenylazothiazole (2) thus obtained, was dried and recrystallized from ethanol. Yield 86%; mp 270-271°C; IR (KBr): $\bar{\nu}$ = 3421, 3270 (NH2), 1639 (CO) cm-1; 1H NMR (DMSO-d_6): δ = 6.90 (s, C_4-H, thiazole), 7.05-7.40 (m, Ar-H), 8.35 (d, NH2) ppm.

2.2.2. Synthesis of 2-Acetylamino-5-Phenylazo-Thiazole (3) (HL)

A mixture of 2-amino-5-phenylzothiazole (2, 0.01 mol) and 10 ml acetic anhydride was heated in an oil bath at 60-65°C for 1 hour.

The reaction mixture was allowed to cool at room temperature and then recrystallized from ethanol to afford the corresponding 2-acetylamino-thiazole derivative (HL).Yield 78%, mp 233-235°C, IR (KBr): $\bar{\nu}$ = 3280 (NH), 1697 (CO), 1562 (C=N) cm-1; 1H NMR (DMSO-d6): δ = 1.80 (s, CH3), 6.85 (s,

C4-H, thiazole), 7.10-7.45 (m, Ar-H), 11.40 (s, NH) ppm. Steps of preparation of ligand are shown in Scheme (1).

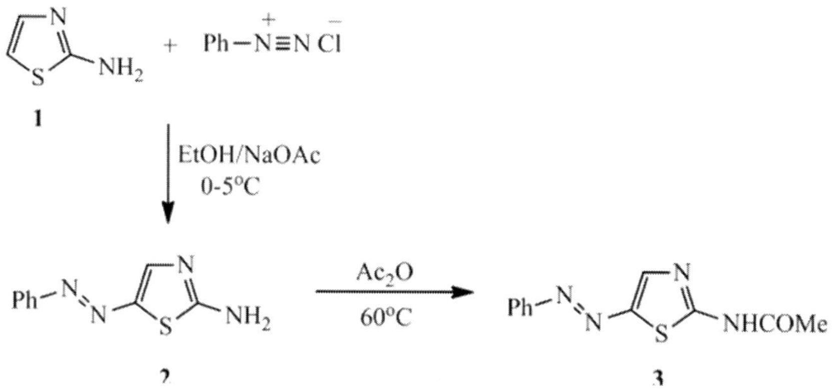

Scheme 1.

2.3. Preparation of Metal Complexes

Cu(II) , Co(.II), Ni(II) and Fe(III) complexes of HL were prepared by adding 0.1 mol (0.246 g) of ligand dissolved in 50 ml of ethanol to 0.1 mol of the appropriate metal salt dissolved in a least volume of ethanol .The reaction mixture is heated under reflux for about 1-2 hrs. The precipitated product was filtered off, washed several times with ethanol and dried under vacuum in presence of anhydrous $CaCl_2$.

2.4. Heterojunctions Preparation

To obtain an Au/org/inorg/Al cell, the sample has been prepared by using a polished p-type Si wafer with (100) orientation and hole concentration of 1.6 $\times 10^{23} m^{-3}$ with thickness of 400 μm as an inorganic semiconductor. In order to remove the native oxide on p-Si surface, the substrate was etched by CP_4 solution (HF: HNO_3: CH_3COOH in ratio 1:6:1) for 10 s, then rinsed with deionized water and isopropyl alcohol and dried in oven. After the drying method all the p-Si pieces were inserted in Edward E306A, England coating unit in order to fabricate the Al-back contact. This was made by depositing a relatively thick film of Al to the bottom of fifteen p-Si substrates; Evaporation of Al is carried out heating the aluminum wire by a tungsten coil under vacuum of 2×10^{-5} Pa. The evaporation rates as well as the film thickness of the evaporated films were controlled by using a quartz crystal monitor FTM6, the deposition rate was controlled at 0.5 nm/s. After the preparation of the Al-back contact, the samples were inserted in a quartz tube to be annealed in oven under 723 K for 3 min under vacuum of 3×10^{-4} Pa.

The prepared powders were dried at 333 K under roughing vacuum for 5 h then dissolved in DMF with concentration of 30 mg/ml. Figure 1(a) shows the variation of colour occupancy of the prepared dyes in the used transparent bottles. The nano-dyes thin films were deposited by spin-coating technique; was spin-coated on the top of p-Si/Al with a speed of 1500 rpm for 45 s.

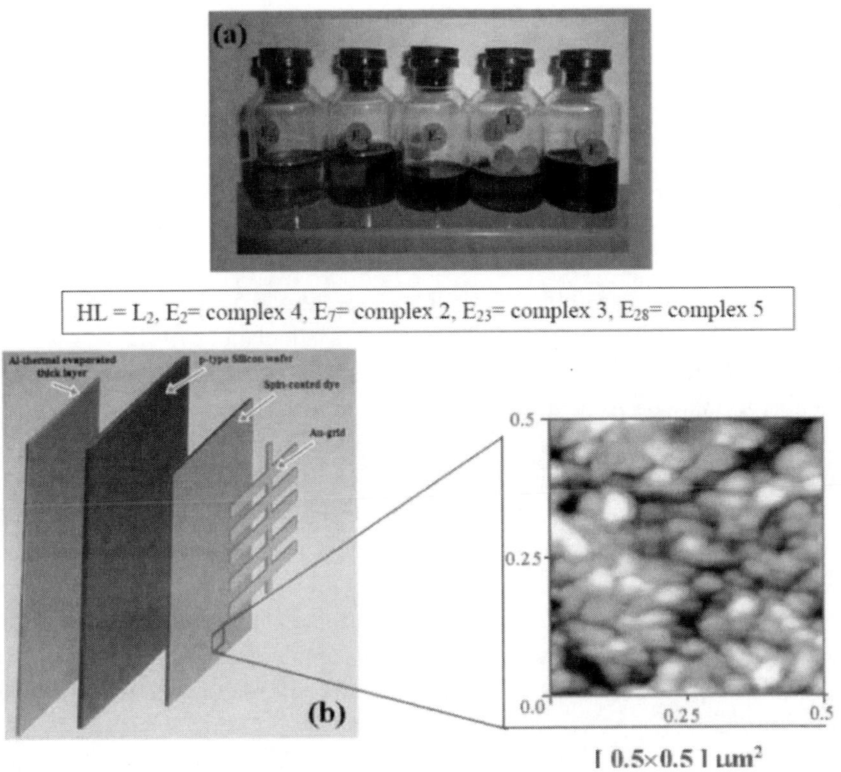

Figure 1. Shows the prepared solutions of the synthesized organic dyes and Org/Inorg-heterojunction structure (a) Suspensions of nearly monodisperse organic dyes, and (b) Schematic structure of the required parts for the Org/Inorg-heterojunctions solar cell devices with gold-mesh. The inset shows the Atomic force microscopy (AFM) image of the E_7-dye, formed with the spin-coating method on p-Si substrate with 0.5×0.5 μm² scan size, as a representative sample.

After finishing the previous step, all samples (HL/p-Si/Al, complex(4)/p-Si/Al, complex(3)/p-Si/Al, complex(5)/p-Si/Al and complex(2)/p-Si/Al) were annealed under roughing vacuum at 353 K; in order to improve the dyeability of the top of p-Si/Al and adhesivity of the dye-film to the top of p-Si/Al. Note, each one of the prepared solution of the prepared dyes was used to prepare a thin film on a three samples of the org/inorg/Al, in order to accurately, investigate the modification of the synthesized dyes for the barrier height between the Au/org and the p-Si/Al of the fabricated heterojunctions. Finally, the Au-wire was thermally evaporated by a tungsten coil under vacuum of 2×10^{-5} Pa to fabricate Au films with an area of 0.06 cm^2 on the spin coated dye-films. The device had the same structure as the previous published works of the fourth author (see Ref. [6]). Figure 1(b) shows a simple diagram for the structure of Au/org-dye or complex:Si-p/Al Org/Inorg heterojunction.

Surface topography was studied by contact mode atomic force microscopy AFM in (Engineering and Surface Metrology Department, National Institute for Standards, Egypt). The collected 3D topographical data was analyzed using data analysis software (IP; Image Processing and Data Analysis, V.2.1.15, Thermo Microscopes, CA, USA) and SPIP software (Image Metrology). Nano-roughness data was calculated over the whole image (area), with cited numbers resulting from at least three images of $0.5 \times 0.5\ \mu m^2$. Plane correction was applied to each image before the roughness calculations [36].

The obtained cell was annealed in air at 353 K for 1 h to complete the junction formation. Annealing of heterojunction has been the usual step in obtaining the best efficiency cells. This annealing might remove any channels, which could be raised during the fabrication [6, 37]. In addition, thermal annealing condition was found very important for improving short circuit current, fill factor, and therefore the efficiency of the device [38]. The dark current–voltage (I–V) characteristics of the fabricated cell were measured simultaneously using a high impedance electrometer (Keithley 6517 B). The dark I–V characteristics were obtained in a complete dark chamber at room temperature or inside a dark furnace in case of measurements at higher temperatures.

The illuminated (I–V) characteristics are carried out by using (Keithley 6517 B) under intensity of light =6 mW/cm^2 provided by a tungsten lamp. The intensity of the incident light was measured by using digital lux-meter (BCHA, model 93408).

3. RESULTS AND DISCUSSION

3.1. Structure Investigation and Spectral Studies of Prepared Ligand (HL)

3.1.1. Elemental Analysis of Ligand (HL)

The prepared organic ligand was purified by recrystallization from ethanol. The obtained results of elemental analysis, which are listed in Table 1, confirm the purity of the prepared organic ligand.

3.1.2. IR Spectrum of ligand (HL)

The IR spectrum of the ligand was recorded in the solid state as potassium bromide discs, within the range 4000-200 cm^{-1}. The assignment of the most diagnostic IR absorption bands were listed in Table 1.

The infrared spectrum of the ligand shows bands at 3159 and 3280 cm^{-1} that assigned to υ(NH) . The spectrum of the ligand also shows strong bands at 1697, 1562, 1435 and 790 cm^{-1} corresponding to υ(C=O), υ(C=N)of the ring , υ(N=N) and υ(C-S) of the ring, respectively.

3.2. Elucidation of Structure of Metal Complexes

3.2.1. Elemental Analysis and Molar Conductivity

The reactions of the ligand HL with different metal ions; $Cu(NO_3)_2$, $CoCl_2$, $NiCl_2$, and $FeCl_3$, in molar ratios 1:1 gave complexes shown in Table 1. As shown in Table 1 and Scheme 2, the reaction of ligand with $Cu(NO_3)_2$and $CoCl_2$ led to the formation of 1:1 molar ratio, whereas, with $NiCl_2$ and $FeCl_3$, 1:2 metal complexes were formed . The obtained metal complexes are air stable, non-hygroscopic and partially soluble in water and most organic solvents except dimethylformamide (DMF) and dimethylsulphoxide (DMSO), where they are freely soluble.

The values of molar conductivities in DMF (10^{-3} M) solutions which are listed in Table 1, indicate that all metal complexes except, complex (5) are non-electrolytes[39], indicating that all anions in metal complexes are directly attached to the metal ion. The value of molar conductivity of complex (5) is 87 $ohm^{-1}cm^2mol^{-1}$ indicating that the solution of the complex is 1:1 electrolyte, and one chloride ion presents in the outer sphere [40].

Table 1. Elemental analysis, color, molar conductivity and molar ratio for ligand and its metal complexes

No	Compounds	color	Λ_M	Salt	Molar ratio	Elemental analysis			
						C %	H %	N %	M%
1	HL	orange	-	-	-	54.0 (53.7)	3.8 (4.1)	22.6 (22.8)	-
2	[CuHL (NO_3)]	brown	20.3	$Cu(NO_3)_2$	1:1	36.1 (35.6)	3.1 (2.7)	18.1 (18.9)	17.5(17.1)
3	[CoHL$Cl_2(H_2O)_2$]	red	19.5	$CoCl_2$	1: 1	32.7 (32.1)	2.9 (3.4)	14.3 (13.6)	14.5(14.3)
4	[Ni$(HL)_2Cl_2$]	green	15.3	$NiCl_2$	1:1	41.9 (42.5)	3.2 (3.2)	17.7 (18.0)	10.0(9.4)
5	[Fe$(HL)_2Cl_2$]Cl	brown	87	$FeCl_3$	1:1	41.0 (40.3)	3.3 (3.1)	17.3 (17.1)	9.1(8.6)

Λ_M = molar conductivity, $ohm^{-1}cm^2mol^{-1}$.

Table 2. Infrared spectral bands for ligand and its metal complexes

No	Compound	υ (N - H)	υ (C= O)	υ (C=N)	υ (N=N)	υ (C- S)	υ (M-S)/ υ (M-O)	υ (M-N) / υ (M-S)
(1)	HL	3159, 3280	1697	1562	1435	790	-	
(2)	[CuL(NO_3)]	-	-	1651, 1580	1470	771	410, 509	455, 410
(3)	[Co(HL)$Cl_2(H_2O)_2$]	3155, 3420[a]	1697	1562	1455	767	410	440, 410
(4)	[Ni$(HL)_2Cl_2$]	3271, 3200	1708	1570	1473	767	420	445, 420
(5)	[Fe$(HL)_2Cl_2$]Cl	3155	1697	1562	1450	768	405	467, 405

[a] = υ (OH) of coordinated water.

3.2.2. Infrared Spectra

The bonding mode of the ligand in the metal complexes has been deduced by comparing the infrared spectra of metal complexes with that of the free ligand. The most diagnostic infrared spectral bands for ligand and its metal complexes and their assignments are listed in Table 2.

The data listed in Table 2 showed that, the infrared spectra of all metal complexes of ligand HL , reveal that the bands ascribed to υ(N=N) at a higher frequency and υ(C-S) band at a lower frequency compared to that of the free ligand, indicating that coordination via the azo group and sulfur atoms occurred.

Table 3. Magnetic moment values and electronic spectral bands of metal complexe in the solid state

No	Complex	μ_{eff} (B.M) per metal ion	d-d bands (nm)
(2)	$[CuL(NO_3]$	1.78	570 (br)
(3)	$[Co(HL)Cl_2(H_2O)]$	4.98	680(w), 580(w)
(4)	$[Ni(HL)_2Cl_2]$	2.98	830, 664
(5)	$[Fe(HL)_2Cl_2]Cl$	5.71	690, 530

br= broad, w= weak.

The spectrum of complex (2) shows that the two bands corresponding to υ(N-H) and υ(C=O) disappeared and a new band appeared at 1651cm^{-1}, assigned to υ(C=N) upon complexation, indicating that the ligand in this complex reacted in the enol form via the enolic oxygen atom , N=N and sulfur atom .The infrared spectra of complexes (3), (4), and (5) show that the bands characteristic to υ(C=N) of the ring and υ(C=O) appeared approximately at the same positions as that of the free ligand, indicating that they don't participate in coordination.

The above arguments together with the results of elemental analyses indicate that, the ligand HL behaves as a monobasic tridentate , coordinating via the sulfur atom, azo nitrogen atom and enolic oxygen atom (NSO mode) or as a neutral bidentate, coordinating through the sulfur atom and azo nitrogen atom (SN mode). The rest coordinations are satisfied by chloride ions, nitrate and /or water molecules to complete the coordination number. The presence of water molecules within the coordinated sphere in the hydrated complexes (3) is supported by the appearance of infrared spectral bands at 3420 cm^{-1}, 885 cm^{-1}, and 695 cm^{-1}, assigned to OH stretching , H_2O rocking and H_2O

wagging, respectively [41]. The spectra of complexes show two new bands at 467 – 440 cm^{-1} and 420 -405 cm^{-1}, assigned to υ(M-N) and υ(M-S), respectively [42]. The spectrum of complex (2) shows additional new band at 509 cm^{-1}, assigned to υ(Cu-O) [42].

3.2.3. Magnetic Moments and Electronic Spectral Data for Metal Complexes

The electronic absorption spectral bands as well as the room temperature magnetic moments (μ_{eff} B.M.) per a metal atom of the metal complexes are listed in Table 4.

The magnetic moment value for the copper(II) complex (3) is 1.78 B.M.. The value of magnetic moment of complex is consistent with the spin only value (1.73 B.M.), correspond to one unpaired electron [43] in a square planar or octahedral geometry [44]. Value also indicates that no exchange interactions occurs between copper ions. The electronic spectrum of copper(II) complex shows a broad band at 570 nm, which suggests a square planar geometry around the copper(II) ion[45].

Table 4. TGA data of some metal complexes

No	Complex	Temperature (°C)	Weight loss found (calcd.) %	Species removed or formed
(2)	[CuL(NO_3)]	30- 164	-	Stable
		164-240	17.2(16.7)	-1 NO_3
		240-872	65.9(66.4)	-1 ligand
		872	-	CuO formed
(3)	[Co(HL)Cl_2(H_2O)$_2$]	30-250	-	stable
		250-305	26.8 (26.0)	- 2 H_2O +2 Cl
		305-951	60.7(59.8)	- ligand
		951	-	CoO formed
(4)	[Ni(HL)$_2Cl_2$]	30-300	-	stable
		300 -900	78.9 (79.8)	- 2 Cl +2 ligand
		900	-	-NiO formed

Octahedral, tetrahedral and square planar cobalt(II) complexes show magnetic moment values between 4.7 – 5.2, 4.2 – 4.8 and 2.2 – 2.9 B.M., respectively [46-48]. The room temperature magnetic moment value for cobalt

(II) complex (3) is 4.98 B.M . This value is indicative for an octahedral geometry around the cobalt(II) ion .The considerably high value ruled out any spin- spin interactions between cobalt(II) ions. The electronic spectrum of cobalt(II) complex shows two weak bands at 680 and 590 nm which may be assigned to $^{4}A_{2}g\rightarrow {}^{4}T_{1g}$ (P) (υ_3) and $^{4}T_{1g}\rightarrow {}^{4}A_{2g}$ (υ_2), respectively, which is indicative for octahedral cobalt(II) geometry [49].

The nickel(II) complex was found to be paramagnetic which excludes the possibility of a square planar configuration. The room temperature magnetic moment value of nickel(II) complex (4) is 2.98 B.M.

Six-coordinate nickel(II) complexes generally exhibit three d–d bands in the Vis.-NIR region corresponding to the $^{3}A_{2g}(F)\rightarrow {}^{3}T_{2g}(F)$ (υ_1), $^{3}A_{2g}(F)\rightarrow {}^{3}T_{1g}(F)$ (υ_2) and $^{3}A_{2g}(F)\rightarrow {}^{3}T_{1g}(p)$ (υ_3) transitions. However nickel(II) complex (4) displays only two bands at 664 and 830 nm, which are attributable to $^{3}A_{2g}(F)\rightarrow {}^{3}T_{1g}(F)$ (υ_2) and $^{3}A_{2g}(F)\rightarrow {}^{3}T_{1g}(p)$ (υ_3) transitions, respectively indicating an octahedral nickel(II) complex. The third one does not observe possibly because it is out the range of spectrophotometer (> 900 nm) [50,51]. Iron(III) complex (5) shows bands at 530 and 690 nm assigned to $_{6}A_{1g}\rightarrow {}_{4}T_{2g}(G)$ and $^{6}A_{1}(G)\rightarrow {}_{4}T_{1}(G)$ transitions, characteristic of octahedral iron(III) [50,52]. The iron(III) complex (5) gave magnetic moment value 5.71 BM, compatible with high-spin d^{5} iron(III) ions.

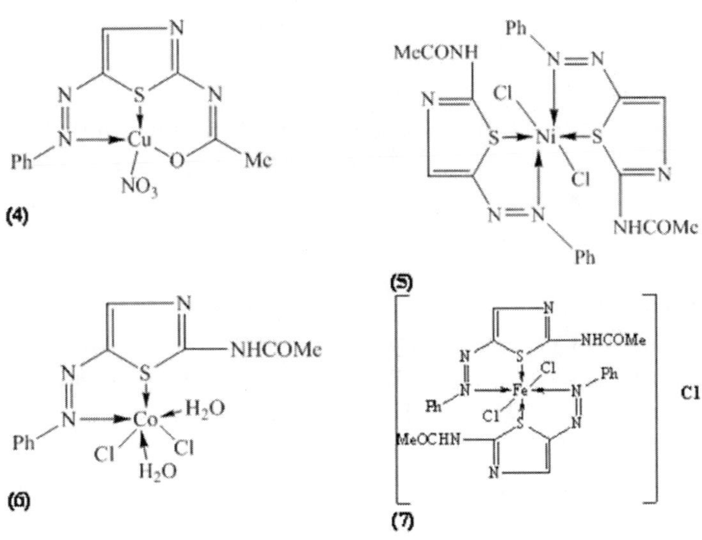

Scheme 2

3.2.4. Thermal Analysis

The TGA results for some of the solid complexes (**2**), (**3**) and (**4**) are listed in Table 4. The results show good agreement with the formulae suggested from the analytical, spectral and magnetic data. A general decomposition pattern was concluded, whereby the complexes decomposed in stages. The first decomposition stage is the loss of the NO_3^- or Cl^- anions at 164-305 °C. The second decomposition stage is the loss of molecules of ligand at temperature range 240 -951 °C for all complexes.

Finally, metal-oxide formation takes place. Complexes (**2**),(3) and (4) show thermal stability up to 164, 250 and 300 °C respectively. The above arguments confirmed the chemical formulas of metal complexes as shown in Scheme 2.

3.3. Effects of Annealing Temperature on Topological Properties of the Prepared Compounds

The inset of Figure 1(b) shows a two-dimensional Topographical image for complex (2) on p-Si/Al as a representative sample. This figure depicts the effects of annealing temperature at 253 K on the surface topography of the spin coated complex (2) synthesized complex as a representative sample. Table 1 shows the recorded parameters which are based on two-dimensional standards.

The values of R_q and $R_{p\text{-}v}$ are decreased by amounts of 1.45 and 8.7 nm, respectively, (may be due to the aggregate densification effect of the annealing temperature 353 K). Also The valley depth (R_v) and peak profile (R_p) are decreased by amounts of 3.38 and 5.3 nm, respectively. In conclusion, the annealing temperature at 353 K decreases the surface roughness R_q, mean peak-to-valley profile roughness $R_{p\text{-}v}$ and mean half wavelength (S_{hw}) of the as-deposited NiTPP thin films by 23.27 %, 21.9 % and 25.17 % respectively.

All analyzed data for the annealed complex (2) spin-coated film on p-Si/Al thin films are limited to the annealing temperature at 353 K in a roughing vacuum. In conclusion, the thermal annealing can modify the surface topographies of the spin-coated films by controlling the aggregate densification and porous properties of the organic dye [37].

3.3. Dark I–V Characteristics of the Prepared Org/Inorg Heterojunctions

Figure 2 shows the typical current–voltage (I–V) characteristic observed for the prepared heterojunctions under the forward and reverse bias measured for all prepared devices and in the voltage range −2 to 2 V.

The device exhibits rectification behaviour whereby the current–voltage (I–V) characteristics are limited by the properties of the inorganic semiconductor substrate and the magnitude of the energy barriers at the hetero-interface.

The low and weak voltage dependence of the reverse bias current as well as the strong voltage dependence of the forward bias current are characteristic properties of strong rectification performance for the junction diode interface; the RR has been estimated to be 1128 at ±1V at room temperature for Au/E_7:Si p-type/AL as a representative example.

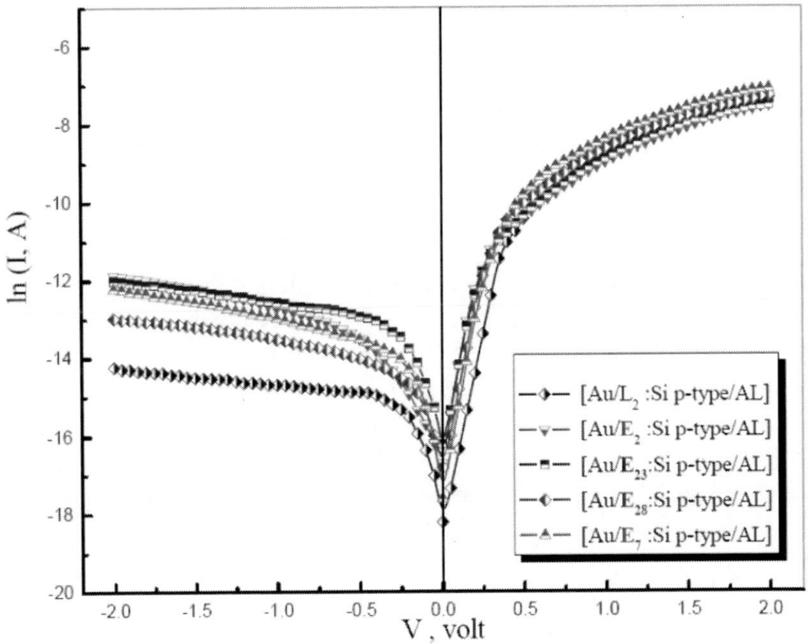

HL = L_2, E_2= complex 4, E_7= complex 2, E_{23}= complex 3, E_{28}= complex 5

Figure 2. (I–V) characteristics for all prepared devices in forward and reverse conditions.

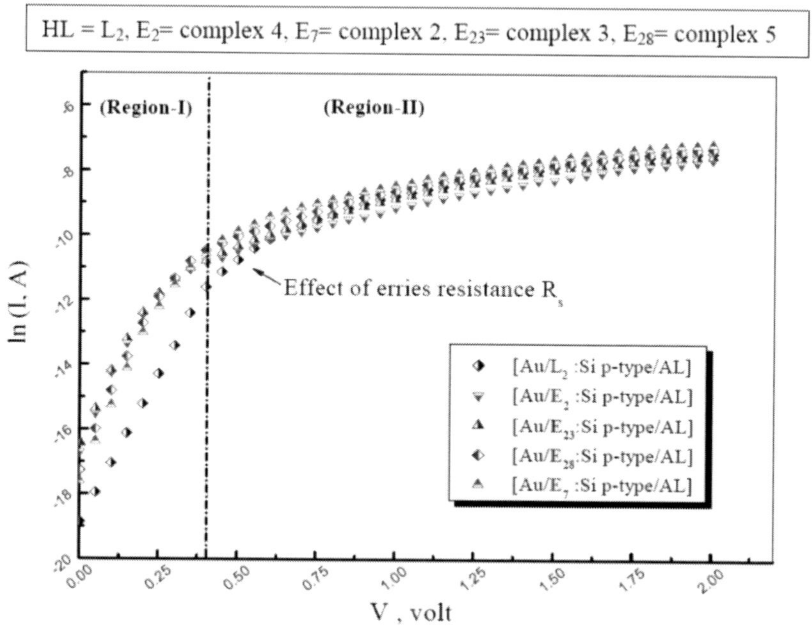

Figure 3. Dark I–V characteristics of the prepared devices at room temperature in forward bias.

The value of the RR is relatively high although the prepared materials are a p-type semiconductor as observed from experiments in our laboratory; this is probably due to the difference in the work functions of p-(synthesized materials) and p-Si as well as the type of conduction in this voltage range [53]. Figure 3 shows the current–voltage (I–V) characteristic observed for the heterojunction under the forward bias for all prepared devices at room temperature. It is obvious that, there is a high series resistance effects at relatively high voltage region $V_F > 0.4$.

When an organic film is inserted between metal and semiconductor, the current–voltage characteristics can be analyzed using the following equation [53, 54]:

$$I = I_o \exp\left(\frac{q(v - IR_s)}{nk_B T}\right)\left[1 - \exp\left(-\frac{q(V - IR_s)}{k_B T}\right)\right], \quad (1)$$

$$I_o = AA^* T^2 \exp\left(\frac{-q\Phi_{bo}}{k_B T}\right), \quad (2)$$

where I_o is the reverse saturation current, q is the elementary charge, V is the applied voltage, n is the diode ideality factor, K_b is the Boltzmann's constant, T is the temperature in Kelvin, R_s is the series resistance, A is the effective area and A^* is the effective Richardson constant that takes the value 32 A/cm^2 K^2 for p-Si [55] and Φ_{bo} is the zero-bias barrier height which is expressed as:

$$\Phi_{bo} = \frac{k_B T}{q} \ln\left(\frac{AA^*T^2}{I_o}\right), \quad (3)$$

The value of the ideality factor, n, is calculated from the slope of the linear portion of forward bias of (I–V) characteristic by using the following equation:

$$n = \frac{q}{k_B T}\left[\frac{dv}{d(\ln I)} - IR_s\right], \quad (4)$$

Under relatively low forward bias condition ($V_F < 0.4$ V), voltage across the series resistance (IR_s) can be neglected, then, Eq. (4) is simplified to the following equation:

$$n = \frac{q}{k_B T}\left[\frac{dv}{d(\ln I)}\right], \quad (5)$$

By using Eqs. (3) and (5), the experimental values of the Φ_{bo} and the ideality factor, n, for each temperature are determined from the intercept and slope of the relation ln(I) vs. V for forward bias $V_F < 0.4$ V. It should be noted that barrier height the, Φ_{bo}, is the contact barrier that exist at the interface between the organic and the inorganic layers. The heterojunction barrier height controls the injection of charge from the metal/organic contact into the inorganic semiconductor substrate [56]. Figure 4 shows the temperature dependence of n and Φ_{bo} for all devices, from which it can be seen that the values of n and Φ_{bo} depends on the spin coated organic material on the p-Si at the temperature room temperature. The lowest value of n is 1.92 and 1.93 for the prepared materials (complexes (2) and (3)), respectively and the highest one is 2.17 for (HL). The lowest value of the modificated barrier height is 0.721 eV and the ideality factor due to this modification is 2.03 for complex(4). The lowest Φ_{bo} was modificated by (complex(3)) and its value is 0.719 eV and the intermediate value for n and Φ_{bo} is 1.97 and 0.734 eV for (complex(5)). The previous investigated results reveal that the prepared dye and its metal complexes can play rules in the modification of the barrier height of the heterojunctions to control the injection of charges from the metal/dye contact into the p-Si/Al in electronic applications [57, 58]. Boyarbay et al. [59] reported that although the standard thermionic emission assumes the barrier height is independent of the temperature; this result confirms that the Reactions of HL with metals have the ability to modificate the barrier height of the fabricated devices and also the noticed different values of the ideality factor may be returned to the nature of the prepared dyes. All the fabricated devices are investigated at the room temperature 293 K and the temperature dependence of the ideality factor and the barrier height not include in our study.

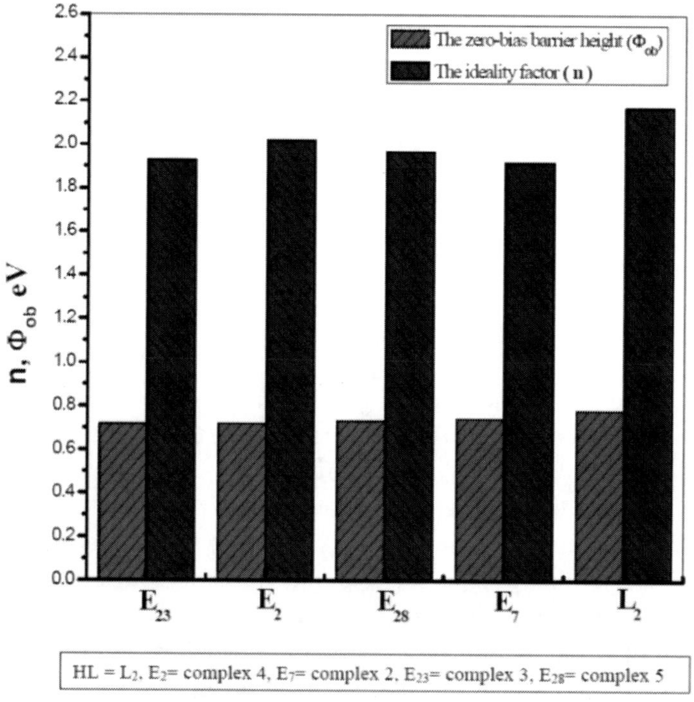

Figure 4. The dependence of the ideality factor, n, and the barrier height, Φ_{ob}, on the nature of the prepared organic material on the p-Si.

In the relative higher voltage region, $V_F > 0.5$ V, the concave deviation from straight line is caused by the presence of the interface states and the series resistance, R_s, associated with bulk Si and the ohmic contacts. The series resistance is a very important parameter of the heterojunction. When the applied voltage is sufficiently large, the ideality factor and the series resistance were evaluated using a method developed by Cheung and Cheung [60]. The Cheung's method is achieved by using Eq. (1) in the following form [55]:

$$V = IR_s + n\Phi_{bo} + n\frac{k_B T}{q}\ln\left(\frac{I}{AA^*T^2}\right), \tag{6}$$

$$\frac{dV}{d(\ln I)} = \frac{nk_B T}{q} + IR_s, \tag{7}$$

$$H(I) = V - n\frac{k_B T}{q}\ln\left(\frac{I}{AA^*T^2}\right), \tag{8}$$

$$H(I) = IR_s + n\Phi_{bo}. \tag{9}$$

Figure 5 shows the plot of [dV/d(ln I)] vs. I. The slope of the linear fit gives the series resistance while its intercept gives the ideality factor at room temperature. The temperature dependence of the series resistance and the ideality factor values are given inset Figure 5. Using Eq. (7), the values of Φ_{bo} and Rs were calculated at different temperatures from the relation between H(I) and I and given inset Figure 6.

Nord proposed an alternative method to determine the series resistance and the barrier height [61]. The following functions have been defined in the modified Nord's method [55, 61] which applied to the full forward bias region of the I–V characteristics of the junction:

$$F(V) = \frac{V}{\gamma} - \frac{k_B T}{q}\left(\frac{I}{AA^*T^2}\right), \tag{10}$$

where γ is the first integer (dimensionless) greater than n which obtained from I–V measurements, here it has been taken as 3. In this function the barrier height can be obtained by using the following equation [55, 61]:

$$\phi_b = F(V_o) + \frac{V_o}{\gamma} - \frac{k_B T}{q}, \tag{11}$$

where $F(V_o)$ is the minimum F(V) value of Nord's function of F(V) vs. graph and V_o is the corresponding voltage. Figure 7 shows the voltage dependence of F(V) at different temperatures. Also, the series resistance, Rs, can be calculated from Nord's functions as [55, 61]:

$$R_s = \frac{k_B T(\gamma - n)}{qI}. \tag{12}$$

The values of the barrier height and the series resistance at different temperatures were obtained using Eqs. (9) and (10) and listed inset Figure 7.

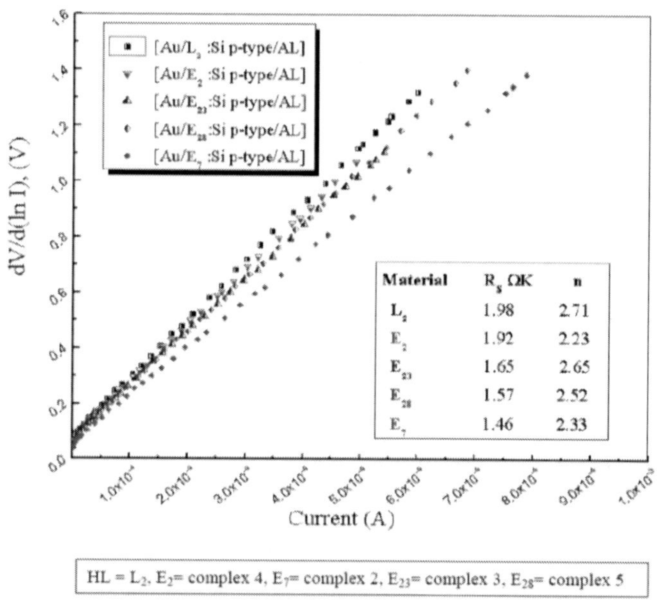

Figure 5. Plot of [dV/d(ln I)] vs. I at room temperature. The inset table indicates the investigated values of the ideality factor, n, and series resistance, R_s, for all fabricated organic dyes/p-Si heterojunctions.

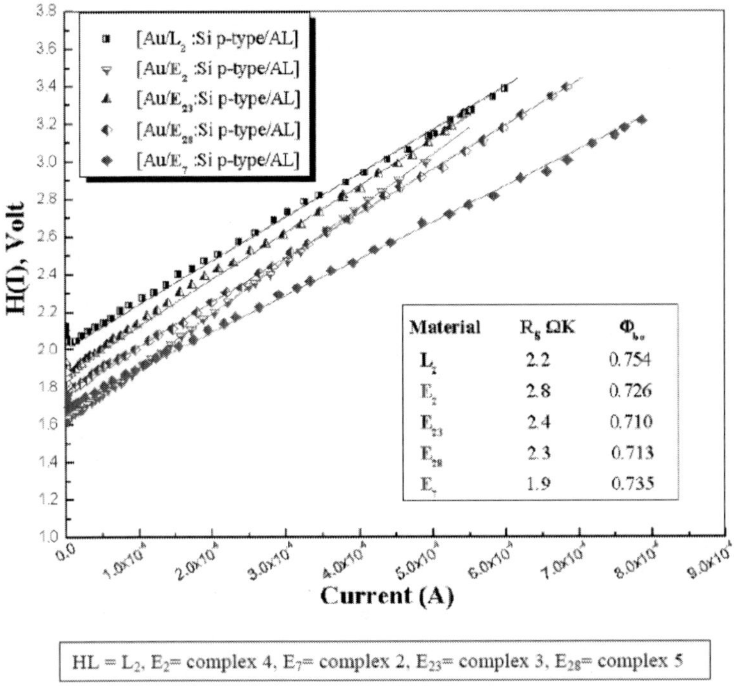

Figure 6. Plot of H(I) and I at room temperature 293 K. The inset table data the investigated values of the barrier height, Φ_{ob}, and series resistance, R_s, for all fabricated organic dyes/p-Si heterojunctions.

The discrepancy between the values of the series resistance and the barrier height obtained from Cheung's functions and Nord's functions may be due to the Cheung's function applied only for the nonlinear region of forward bias while Nord's function are applied for the whole forward bias region of I–V curve [55].

3.4. Photovoltaic Response of the Au/Synthesized-Dyes/P-Si/Al Device

All the prepared heterojunctions by using the pre-prepared organic dyes showed an initial response under illumination. This response depends on the inserted organic layer between the (Au) as an ohmic contact for the inserted organic layer (HL, complex(4), complex(3), complex(5) or complex(2)) and the p-Si as an inorganic semiconductor, the injection of charge is controlled by the inserted organic layer as the I–V characteristic curves (in dark and under illumination) reveals in Figure 8. All the prepared devices have the same active area (0.06 cm^2) and investigated under the same conditions of illumination and with the white light of power 6 mW/cm^2. For each device, it can be seen that at a certain voltage, the current value for each cell under illumination is higher than in the dark condition; this phenomenon is a result of the difference in the electron affinities of the two semiconductors [62, 63].

All (I–V) characteristic curves of maximum power region for the prepared heterojunctions with the different synthesized organic dyes are collected in one master diagram as shown in Figure 9(a). On illumination, the light generates a photocurrent and its influence appears in the region of the fourth quadratic, which is plotted separately in the previous figure.

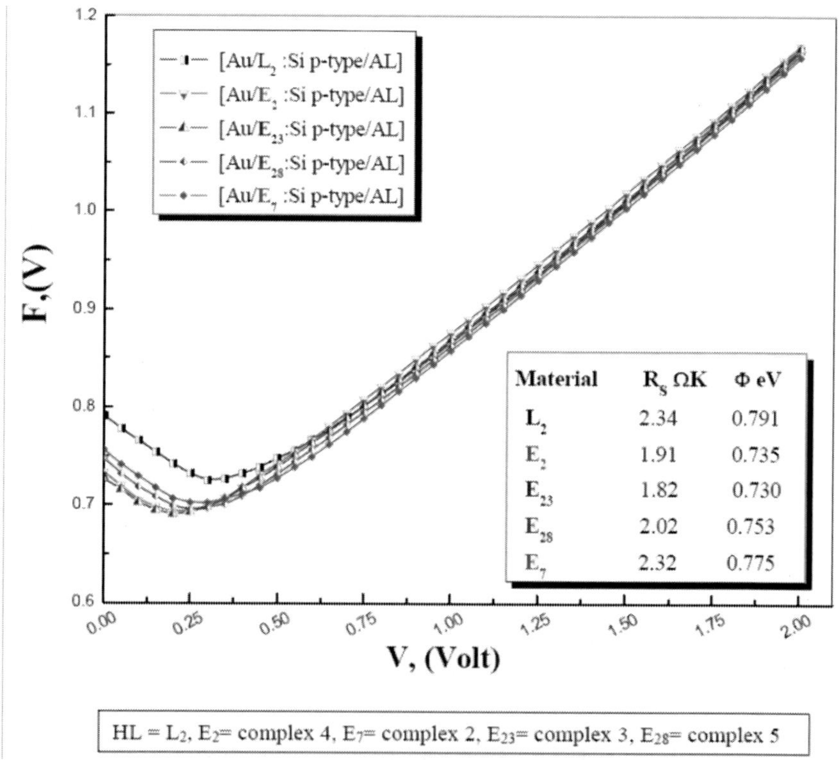

Figure 7. Voltage dependence of F(V) at room temperature 293. The inset table data the investigated values of the barrier height and series resistance by using Nord's method for all fabricated organic dyes/p-Si heterojunctions.

Table 5. Summary of surface morphology parameters from AFM image analyses

Roughness Parameter	Complex(2) thin film category	
	Spin-coated (2) on p/Si/Al at 293 K	Annealed (2)/p-Si/Al at 253 K
R_q (nm) Root Mean Square (RMS) parameter	6.23	4.78
$R_{(p-v)}$ (nm) Mean peak-to-valley profile roughness	39.7	31
R_v (nm) Maximum Valley Depth	12.24	8.86
R_p (nm) Maximum Peak Height	27.43	22.13
Projected area (μm^2)	0.25	0.25
S_{hw} (nm) Mean Half Wavelength	262.2	196.2

All the prepared devices exhibit a short circuit current, I_{sc}, and an open-circuit voltage, V_{oc}. The variation in the resultant values of (I_{sc}) and (V_{oc}) confirms the modification of the charge injection from the metal/organic into the p-Si substrate and depicts that the barrier height is controlled by the inserted organic layer under the same conditions of preparation and measurements. The photovoltaic parameters for each device are investigated and recorded in Table 2. The photoresponsivity value for the heterojunction is defined as the ratio of photocurrent to dark current (I_{photo}/I_{dark}) [63].

As seen in the previous figure, the photoresponsivity of the heterojunction under light illumination due to the photogeneration of charges carriers. The prepared heterojunctions exhibit a different photosensitivity under visible light illumination of 6 mW/cm^2. It is observed that the photocurrent in the device in reverse direction is strongly enhanced by photo-illumination.

This behavior yields useful information on the electron–hole pairs, which were effectively generated in the junction by incident photons. Under the influence of the electric field at the junction, the free electrons and holes were accelerated towards the Au and Al electrodes along the potential barrier at the interface [64]. Figure 9(b) is introduced as a representative sample (Au/complex(2):Si p-type/AL) of the best resultant device of the maximum power region.

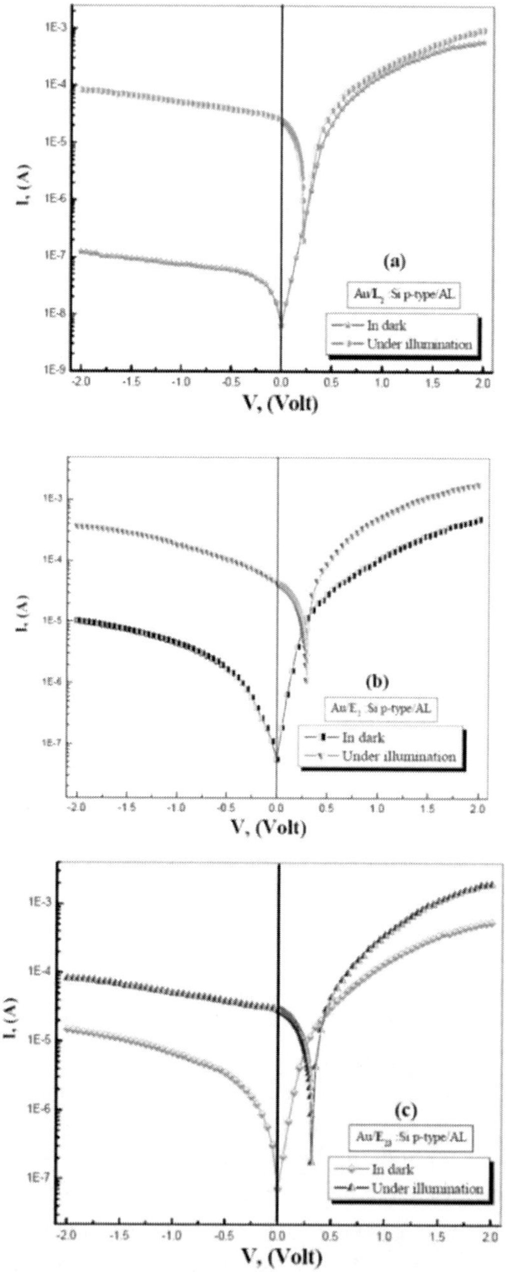

(a)
In dark
Under illumination
I, (A)
V, (Volt)
(b)
In dark
Under illumination
I, (A)
V, (Volt)
(c)
In dark
Under illumination
I, (A)
V, (Volt)

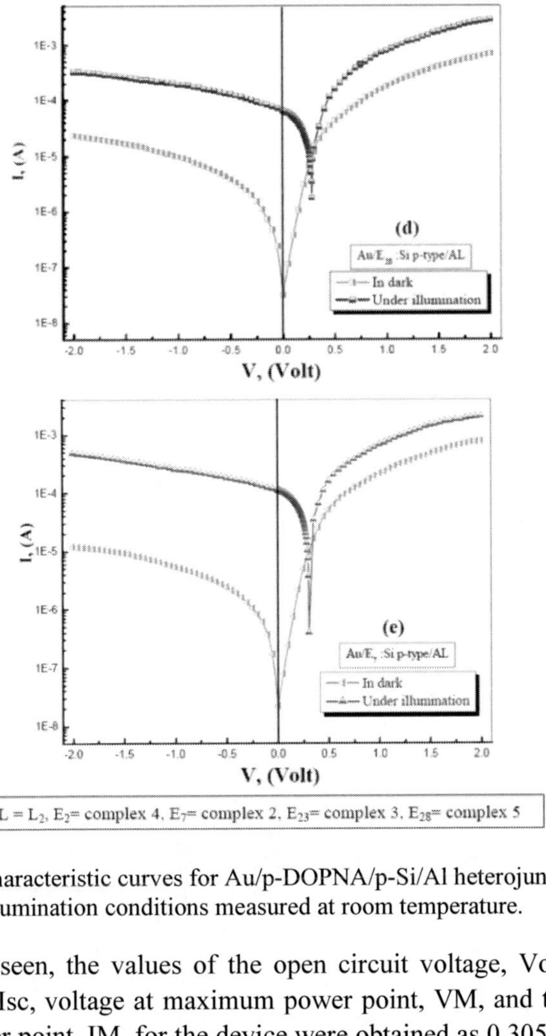

Figure 8. (I–V) characteristic curves for Au/p-DOPNA/p-Si/Al heterojunction diode under dark and illumination conditions measured at room temperature.

As can be seen, the values of the open circuit voltage, Voc, the short-circuit current, Isc, voltage at maximum power point, VM, and the current at maximum power point, IM, for the device were obtained as 0.305 V, 0.11 mA, 0.175 V and 0.068 mA, respectively. The FF is given by [65]

$$FF = V_M I_M / V_{oc} I_{sc} \quad (13)$$

The value of the FF was calculated as 0.35. The experimental power conversion efficiency of a solar cell is given by [65]

$$\eta = (FF V_{oc} I_{sc}) / AP_{in} \times 100\ \% \quad (14)$$

where P_{in} is the illumination intensity impinging on the cell and A the effective area. The efficiency of the cell was calculated as 3.3 %. The fill factor value of complex(2)/p-Si device, like some other organic based devices, is lower in comparison to those of solar cells based on inorganic materials.

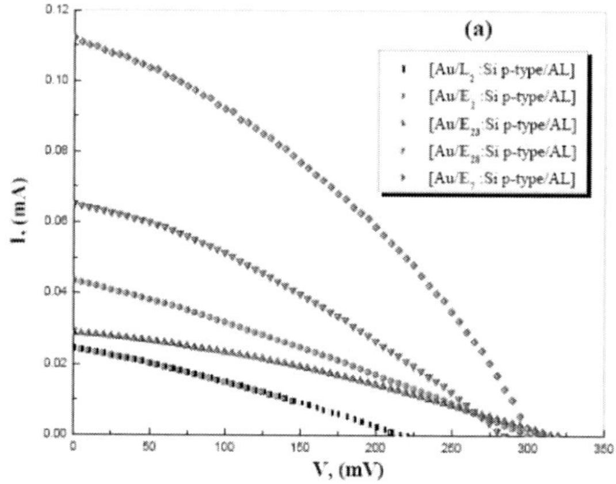

HL = L_2, E_2= complex 4, E_7= complex 2, E_{23}= complex 3, E_{28}= complex 5

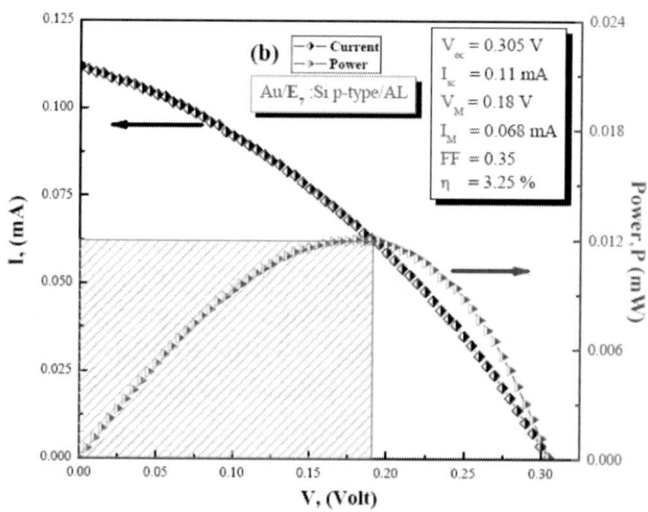

Figure 9. Loaded I–V characteristic and power of (a) all fabricated Au/org-dye:p-Si/Al heterojunction and (b) Au/E_7:p-Si/Al heterojunction under illumination as a representative sample.

The main reason of this result can be attributed to the field dependent nature of the charge photogeneration process or high series resistance of the organic layer [7]. Therefore, it can be said that the complex(2)/p-Si heterojunction has a good photovoltaic parameters and 2-Aminothiazoles and their derivative-dyes which may be alternative candidate for photovoltaic application based on organic molecules. Moreover, as observed from the above comparative study, the presented devices exhibit considerably acceptable photovoltaic properties valid for photovoltaic sensor applications, especially when the reproducibility phenomenon was taken into consideration.

The low hole mobility, the disordered dissociation and charge dissipated recombination in organic materials lead to both lower efficiency and fill factor [66].

Table 6. Photovoltaic parameters of some organic ligand and complex heterojunction solar cells

Solar cell construction	J_{sc} (mA/cm^2)	V_{oc} (V)	Fill factor	Efficiency (%)	Ref.
Au/HL :Si p-type/AL	4.1×10^{-4}	0.215	0.29	4.26×10^{-4}	Present Work
Au/(4) :Si p-type/AL	7.3×10^{-4}	0.3	0.28	1.02×10^{-3}	
Au/(3):Si p-type/AL	0.48	0.315	0.33	0.83	
Au/(5):Si p-type/AL	1.1×10^{-3}	0.285	0.32	1.65×10^{-3}	
Au/(2) :Si p-type/AL	1.86	0.305	0.35	3.3	
Al/TPP/Au	–	0.74	0.18	0.06	[67]
Au/Ni-phthalocyanine/p-Si/Al	3.67	0.35	0.4	1.05	[68]
Ag/p-CuPc/n-GaAs/Ag	0.072	0.20	0.22	2.9×10^{-3}	[69]
pentacene/n-Si photovoltaic device	0.471	0.22	0.42	0.044	[70]
Au/PYR(G)/p-Si/Al	0.72	0.50	0.41	1.44	[7]

CONCLUSION

2-Acetylamino-5-phenylazo-thiazole (HL) and its complexes with Cu(II), Co(II), Ni(II) and Fe(III) have been prepared and characterized . The obtained data show that the reaction of ligand with metal ions led to the formation of 1:1 metal complexes as in case of Cu(II) and Co(II) ,whereas, in case of Ni(II) and Fe(III) 1:2 (metal: ligand) complexes were formed. The results also show that in Cu(II) complex, the ligand behaves as a monobasic tridentate ligand coordinating via the azo nitrogen atom, sulfur atom and the enolic oxygen atom (NSO mode).Whereas, in case of complexes of Co(II), Ni(II) and Fe(III), the ligand behaves as a neutral bidentate, coordinating through the azo nitrogen and sulfur atoms (NS mode).

The analysis of the AFM topological image showed a large scale of nano particles. The current–voltage (I–V) characteristics of the Au/synthesized-dye/p- Si/Al heterojunction have been measured at the room temperature 293 °K. The effect of the synthesized dye and its complexes on the heterojunction parameters as Φ_{ob}, n and R_s, were determined from I–V measurements with considering the inhomogeniety due to the mismatch through the spin-coated dye/p-Si interface. The synthesized dye and its metal complexes modified the ideality factors and barrier heights at room temperature. The dominant operating mechanism in the reverse bias is generation recombination of carriers, which is limited to the process of powder, spin-coated dye on the p-Si and annealing temperature at 353 °K. The investigation of (I–V) characteristics under illumination of =6 mW/cm^2 confirms that the short-circuit current, I_{sc}, is open-circuit voltage, V_{oc}, and the fill factor, FF, are depending on the synthesized dyes. From the extracted experimental data, The fabricated device; Au/complex (2):p-Si/Al is the best one as photovoltaic device with conversion efficiency equal to 3.3 %. Its short-circuit current, I_{sc}, is open-circuit voltage, V_{oc}, and the fill factor, FF, are 0.11 mA, 0.305 V and 0.35%, respectively. The obtained results have suggested the fabricated heterojunction is a potential candidate for photovoltaic sensor application.

Acknowledgments

Thanks are due to Scientific Research Deanship, Qassim University, KSA, for financial supporting this work (Research project number 798).

The author do their appreciation to professors; M.M. El-Nahass and H.M. Abd El-Khalek, for their help for organic/inorganic heterojunction part.

References

[1] Y.-J. Liu and H.-Z. Yu, *J. Phys. Chem.* B. 107 (2003) 7803.

[2] H. Haick, M. Ambrico, T. Ligonzo, R.T. Tung, D. Cahen and *J. Am. Chem. Soc.* 128 (2006) 6854.

[3] Aviram and M.A. Ratner (Eds.), Molecular Electronics: Science and Technology, *Ann. N.Y. Acad. Sci.*, (1998).

[4] Vilan, A. Shanzer and D. Cahen, *Nature* 404 (2000) 166.

[5] M.M. El-Nahass, A.F. El-Deeb, H.S. Metwally and A.M. Hassanien, *Materials Chemistry and Physics 125* (2011) 247.

[6] M.M. El-Nahass, H.S. Metwally, H.E.A. El-Sayed and A.M. Hassanien, *Synthetic Metals* 161 (2011) 2253.

[7] A.A.M. Farag, H.S. Soliman and A.A. *Atta, Synthetic Metals* 161 (2012) 2759.

[8] M.E. Aydin, A.A.M. Farag, M. Abdel-Rafea, A.H. Ammar and F. Yakuphanoglu, *Synthetic Metals 161* (2012) 2700.

[9] El Amrani, B. Lucas and B. Ratier, *Synthetic Metals* 161 (2012) 2566.

[10] L. G. Kaake, P. F. Barbara, and X.-Y. Zhu, *J. Phys. Chem. Lett.* 1(2010) 628.

[11] D.Y. Kim, J. Subbiah, G. Sarasqueta, F. So, H. Ding, Irfan and Y. Gao, *Appl. Phys. Lett.* 95 (2009) 093304-1–093304-3.

[12] Friedel, P.E. Keivanidis, T.J.K. Brenner, A. Abrusci, C.R. McNeill, R.H. Friend and N.C. Greenham, *Macromolecules* 42 (2009) 6741.

[13] H. Beyer, Organic Chemistry, Verlag Harry Deutsch, *Frankfurt-um-Main and Zürich*, 1963, pp. 609.

[14] A.-C. Gaumont, M. Gulea and J. Levillain, *Chem. Rev*. 109 (2009) 1371.

[15] U.G. Ibatullin, T.F. Petrushina, L.Y. Leitis, I.Z. Minibaev, B.O. Logvin, Khim. Geterotsitl. *Soedin*. (1993) 715.

[16] P.C. Hang and J.F. Honek, Bioorg. *Med. Chem. Lett.* 15 (2005) 1471.

[17] P. Beuchet, M. Varache-Lembège, A. Neveu, J.-M. Léger, J. Vercauteren, S. Larrouture, G. Deffieux, A. Nuhrich, *Eur. J. Med. Chem.* 34 (1999) 773.

[18] Geronikaki, P. Vicini, N. Dabarakis, A. Lagunin, V. Poroikov, J. Dearden, H. Modarresi, M. Hewitt and G. Theophilidis, *Eur. J. Med. Chem*. 44 (2009) 473.

[19] Papadopoulou, A. Geronikaki and D. Hadjipavlou-Litina, *Farmaco* 60 (2005) 969.

[20] Kreutzberger and H. Schimmelpfennig, *Arch. Pharm*. 314 (1981) 385.

[21] Keil, R. Flaig, A. Schroeder and H. Hartmann, Dyes Pigments 50 (1) (2001) 67.

[22] M.A. Metwally, E. Abdel-latif, A.M. Khalil, F.A. Amer and G. Kaupp, *Dyes Pigments* 62 (2) (2004) 181.

[23] K. Singh, S. Singh and J.A. Taylor, *Dyes Pigments* 54 (3) (2002) 189.

[24] M.S. Yen and I.J. Wang, *Dyes Pigments* 67 (3) (2005) 183.

[25] Y. Geng, D. Gu, F. Gan, *Optical Materials,* 27 (2004) 193.

[26] W. Bin, W. Yi-Qun, G. Dong-Hong, G. Fu-Xi, *Chinese Physics Letters*, 20 (2003) 1596.

[27] H. Fu-Xin, W. Yi-Qun, G.Dong-Hong, G. Fu-Xi, *Chinese Physics Letters,* 20 (2003) 2259.

[28] E. Hamada, T. Fujii, Y. Tomizawa, S. Limura, *Japanese Journal of Applied Physics*, 36 (1997) 593.

[29] Y. Suzuki, Y. Okamoto, Y. Kurose, S. Maeda, *Japanese Journal of Applied Physics,* 38 (1999) 1669.

[30] E. Radziemska, *Energy Convers. Manag*. 46 (2005) 1485.

[31] V. Badescu, *Energy Convers. Manag*. 47 (2006) 1146.

[32] K. Bouzidi, M. Chegaar and A. Bouhemadou, *Sol. Energy Mater. Sol.* Cell 91 (2007) 1647.

[33] Z. Zhou, K. Zhao, *Energy Convers. Manag*. 52 (2011) 2153.

[34] M. Ramadan, I. M. EL-Mehasseb. *Trans. Met. Chem*. 23 (1998) 183.

[35] E. W. Ainscough, A. M. Brodie, A. J. Dobbs, J. D. Ranford, J. M. Waters, *Inorg. Chem. Acta* 267 (1998) 27.

[36] M.M. El-Nahass, H.M. Abd El-Khalek and Ahmed M. Nawar, *Optics Communications,* 285 (2012)1872.

[37] M.M. El-Nahass, H.M. Abd El-Khalek and Ahmed M. Nawar, *Eur. Phys. J. Appl. Phys.* (2012) DOI: 10.1051/epjap/2012110280.

[38] M.M. El-Nahass, A. A. Atta, H. E. A. El-Sayed and E. F. M. El-Zaidia, current organic chemistry, *ISSN*:13852728(2010)1.

[39] W. J. Geaey, *Coord. Chem. Rev*.7 (1971) 81.

[40] N.N. Greenwood, B.P. Straughan, A.E. Wilson, *J. Chem. Soc. A* (1968) 2209.

[41] A.S. El-Tabl, F.A. El-Saied and A.N. Al-Hakimi ,*Transition Met. Chem.* ,32 (2007) 689

[42] M. Teotia, J.N. Gurthu, V.B. Rama, *J. Inorg. Nucl. Chem*. 42 (1980) 821

[43] K.B. Gudasi, S.A. Patel, R.S. Vadvavi, R.V. Shenoy, M. Nethaji, *Trans. Met. Chem.,* 586 (2006) 31.

[44] El-Motaleb, M. Ramada, W. Sawodny, H.F. El-Baradie, M. Gaber, *Trans. Met. Chem*., 211 (1997) 22.

[45] P. Lever, Inorganic Electronic Spectroscopy, Second edition, *Elsevier Science Publishing Company*, Amsterdam, 1984.

[46] Labadi, L. Horvath, G. Liptay, *J. Therm. Anal. Cal*. 83, (1) (2006) 247.

[47] Cukurovali, I. Yilmaz, *Trans. Met. Chem.* 31 (2006) 207.

[48] D.X. West, L.K. Pannell, *Trans. Met. Chem*. 14 (1989) 457.

[49] P.K. Singh, D.N. Kumar, *Spectrochim. Acta, Part A* 64 (2006) 853.

[50] B. P. Lever “Inorganic Electronic Spectroscopy” *El-Sevier Amsterdam* (1968).

[51] M.A. Ali, A.H. Mirza, C.Y. Yee, H. Rahgeni, P.V. Bernhardt; *Polyhedron* 30 (2011) 542–548
[52] A.S. El-Tabl, M.M. E. Shakdofa, A. M.A. El-Seidy, Ahmed N. Al-Hakimi, *J. Korean Chem. Soc.* 2011, Vol. 55,19-27.
[53] K.R. Rajesh and C.S. Menon, *J. Non-Cryst. Solids* 353 (2007) 398.
[54] S.M. Sze and K.K. Ng, Physics of Semiconductors Devices, 3rd ed., *John Wiley and Sons, Inc.*, Hoboken, NJ, 2007.
[55] N. Kavasoglu, C. Tozlu, O. Pakma, A.S. Kavasoglu, S. Ozden, B. Metin, O. Birgi and S. Oktik, *Synthetic Met.* 159 (2009) 1880.
[56] Temirci, M. C¸ akar, A. Türüt and Y. Onganer, *Phys. Status Solidi* (a) 201 (2004) 3077.
[57] Dietrich R.T. Zahn, Sunggook Park and Thorsten U. Kampen, *Vacuum* 67 (2002) 101.
[58] M. Fahlman and W.R. Salaneck, *Surface Science* 500 (2002) 904.
[59] Boyarbay, H. C¸ etin, A. Uygun and E. Ayyildiz, *Thin Solid Films* 518 (2010) 2216.
[60] S.K. Cheung and N.W. Cheung, *Appl. Phys. Lett.* 49 (1986) 85.
[61] H. Nord, *J. Appl. Phys.* 50 (1979) 5052.
[62] M.M. El-Nahass, A.M. Farid and A.A.M. Farag, H.A.M. Ali, *Vacuum* 81 (2006) 8.
[63] F. Yakuphanoglua and W.A. Farooq, *Synth. Met.* 161 (2011) 379.
[64] G.D. Sharma, S.K. Gupta and M.S. Roy, *Thin Solid Films* 333 (1998) 176.
[65] H.M. Zeyada, M.M.El-Nahass and E.M.El-Menyawy, *Solar Energy Materials and Solar Cells* 92 (2008) 1586.
[66] W. Hu, M. Matsumura, *J. Phys., D Appl. Phys.* 37 (2004) 1434.
[67] K. Takahashi, S. Nakatani, T. Yamaguchi, T. Komura, S. Ito, K. Murata, *Sol. Energy Mater. Sol. Cell* 45 (1997) 12.
[68] S. Riad, *Thin Solid Films* 370 (2000) 253.
[69] Kh.S. Karimov, M.M. Ahmed, S.A. Moiz, M.I. Fedorov, *Sol. Energy Mater. Sol. Cells* 87 (2005) 61.
[70] N. Oyama, Y. Takanashi, S. Kaneko, K. Momiyama, K. Suzuki and F. Hirose, *Microelectronic Engineering* 88 (2011) 2959.

In: Smart Nanoobjects
Editor: Kirill Levine

ISBN: 978-1-62808-741-3

Chapter 3

Synthesis and Optical Properties of Rare-Earth and Transition Metals Doped Titanium Dioxide Films

Y. V. Gerasimenko*[*]*, V. A. Logacheva and A. M. Khoviv
Voronezh State University, Voronezh, Russia

Abstract

In this paper we report the formation and investigation of a series of titanium dioxide films doped by rare-earth and transition metals. $M/TiO_2/Si$(or quartz) (M = La, Sm, Ni, Co) two-layers structures were prepared using magnetron and high-frequency plasma sputtering. Subsequent thermal treatment and pulse photon annealing resulted in the formation of compound oxide films consisting of $La_2(TiO_3)_3$ and La_2TiO_5, Sm_2TiO_5, $NiTiO_3$, $CoTiO_3$. Comparing with initial TiO_2 short-wave transmission, the edge of energy band shifted. La and Sm dopants shift this edge to short-wave region, while Ni and Co dopant - to long-wave region. The corresponding analysis of edge absorption and transition energy values determination is presented.

Keywords: Thin films; titanium dioxide; compound oxide; doping; magnetron sputtering; high-frequency plasma sputtering; oxidation; absorption spectroscopy

[*] E-mail: yuliya-gerasimenko@yandex.ru

INTRODUCTION

Thin-film optical coatings are regarded as potential surface materials for science, power engineering, medicine, etc. One of the most significant scopes of antireflection coatings is the solar power engineering [1, 2]. Principal factor influencing coatings quality and in the case of vacuum technological processes is the factor of a composition and a quality of the film-forming materials used for deposition of optical coatings [3]. The list of materials applicable to manufacture of thin-film coatings is large. Equally with single-component materials (TiO_2, ZrO_2, HO_2, SiO_2) the increasing application is found by mixes of substances and solid solutions on their basis [4].

New composite materials based on titanium dioxide are of the greatest interest. Titanium dioxide films are used in photovoltaics as antireflection coatings due to their excellent optical properties, chemical resistance, thermal stability [5, 6]. Doping of rare-earth and transition metals allow to improve optical properties, mechanical and exploitation stability of TiO_2 [7].

In addition, rare-earth and transition metals doping is required for luminescent properties. The luminescence of optical coatings increases the efficiency of photovoltaics.

EXPERIMENTAL DETAILS

Doped titanium dioxide nanocrystalline films (M:TiO_2; M = La, Sm, Co, Ni) were successfully synthesized by thermal oxidation and pulse photon annealing of two-layers structures: La/TiO_2/Si(quartz), Sm/TiO_2/Si(quartz), Ni/TiO_2/Si(quartz), Co/TiO_2/Si(quartz) on silica and quartz substrates.

TiO_2 has been deposited by two different techniques, including magnetron sputtering with subsequent thermal oxidation at 723 K and high-frequency plasma sputtering. The next step was the magnetron sputtering of doping metal (M = La, Sm, Ni, Co) with subsequent thermal oxidation at following temperatures: T = 723, 873 and 1023 K and pulse photon annealing. The result of this synthesis was formation of the compound oxide films with thickness ~ 300 nm.

Metal films were deposited on quartz and silicon (100) substrates by magnetron sputtering from a metallic targets (Ti, La, Sm, Ni, Co) (99.8 % purity) in a vacuum chamber (pressure $6 \cdot 10^{-6}$ Torr). Experimental conditions were as follows: the deposition occurred in atmosphere of high purity (99.9997 %) argon (partial pressure $2 \cdot 10^{-3}$ Torr), current 0.5 A for Ti, 0.2 A for La and

Sm, 0.5 A for Ni and 0.7 for Co. Thickness were determined by the deposition time. Titanium dioxide films were also deposited by reactive r. f. sputtering technique from a metallic titanium target (99.8 % purity) in atmosphere of high purity (99.98 %) oxygen as a reactive gas (partial pressure $4.5 \cdot 10^{-2}$ Torr), and argon (99.9997 % purity) as a sputtering gas (partial pressure $9.5 \cdot 10^{-2}$ Torr). Monocrystalline silicon and quartz were used as the substrates and their temperature during deposition was maintained at 400 K. Thickness were determined by the deposition time.

The crystalline structures of the films were measured using X-ray diffraction (XRD) with Cu Kα source (Thermo Scientific ARL X'TRA with ICDD data). SEM (Jeol JSM-6510LV) was used for the surface investigations and thickness measurements of the films. Optical transmittance of films was measured at room temperature with Shimadzu UV-2550 spectrophotometer, using the substrate properties as background references. The photoluminescence emission spectra were recorded with spectral luminescent complex using laser (λ = 337 nm) for photoluminescence excitation at the liquid-nitrogen temperature.

Results and Discussion

It has been established that during thermal treatment process of two-layers structures $M/TiO_2/Si$(quartz) the compound oxide films with thickness 300 nm consisting of $La_2(TiO_3)_3$ and La_2TiO_5, Sm_2TiO_5, $NiTiO_3$, $CoTiO_3$ respectively for each structure has been generated. X-ray diffraction patterns and SEM images of these structures are shown in Figure 1 – Figure 4.

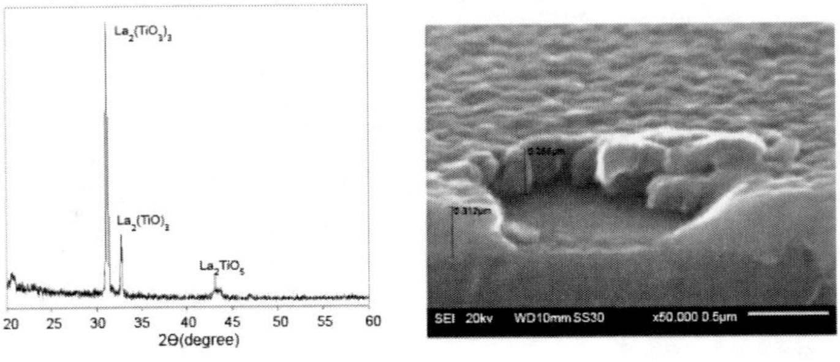

a b

Figure 1. X-ray diffraction pattern (a) and SEM image (b) of the La/TiO$_2$/Si structure after pulse photon annealing.

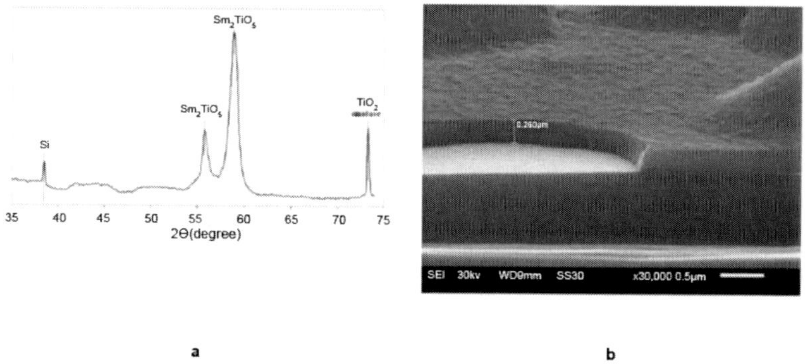

Figure 2. X-ray diffraction pattern (a) and SEM image (b) of the $Sm/TiO_2/Si$ structure after thermal oxidation at T = 1023 K.

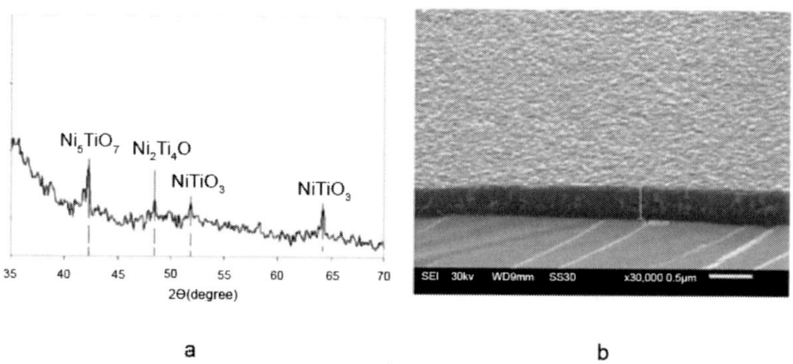

Figure 3. X-ray diffraction pattern (a) and SEM image (b) of the $Ni/TiO_2/Si$ structure after thermal oxidation at T = 1023 K.

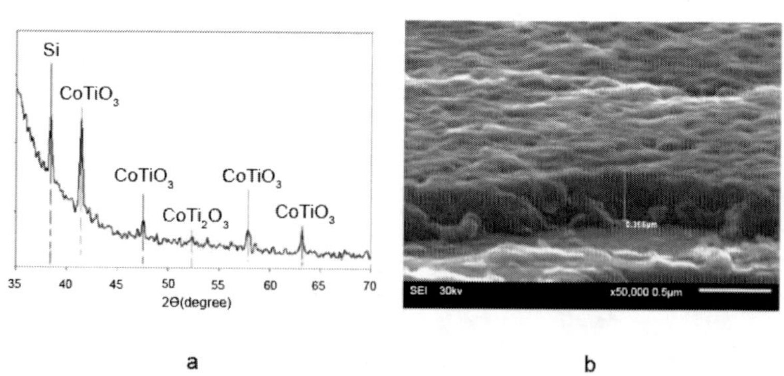

Figure 4. X-ray diffraction pattern (a) and SEM image (b) of the $Co/TiO_2/Si$ structure after thermal oxidation at T = 1023 K.

The transmittance of films deposited on clear quartz was measured over the range of 200 to 900 nm. Figure 5 shows the optical transmittance spectra of thin films TiO_2/quartz and *M*:TiO_2/quartz, *M* = La, Sm, Co, Ni. Absorption spectroscopy shows that TiO_2 film has high transparency in the wavelength range from 350 to 900 nm and absorption band at spectral range from 190 to 350 nm. The transmittance of the TiO_2 film was about 90 %.

Metal used for doping determines the shift of short-wave transmission band edge and the optical transition energy. The films transmission increases through the doping of impurity atoms. For TiO_2 the short-wave edge wavelength is equal to 350 nm, La and Sm doping shift this edge to short-wave region and the edge wavelength becomes 330 and 270 nm respectively. Ni and Co doping shifts this edge to long-wave region and the edge wavelength es to 355 and 370 nm respectively.

The optical band gap energy for the thin films was determined from the sharply falling transmission region, using Tauc's low [8]. The analysis of edge absorption has shown that transition energy value is 3.67 eV for TiO_2. The optical transition energy was 3.70 and 3.06 eV and correspond to $La_2(TiO_3)_3$ and La_2TiO_5 respectively, 4.7 eV for Sm_2TiO_5, 3.64 eV for $NiTiO_3$ and 3.35 eV for $CoTiO_3$.

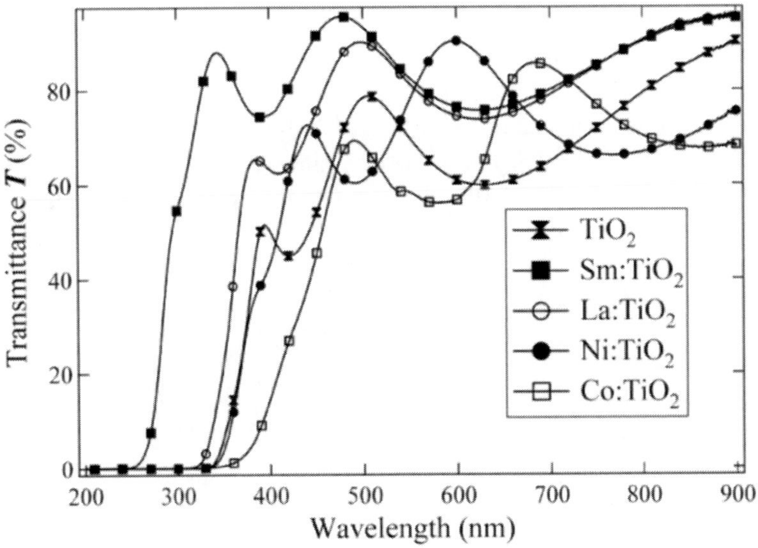

Figure 5. Optical transmittance T spectra of thin films TiO2/quartz and M:TiO2/quartz, M = La, Sm, Ni, Co.

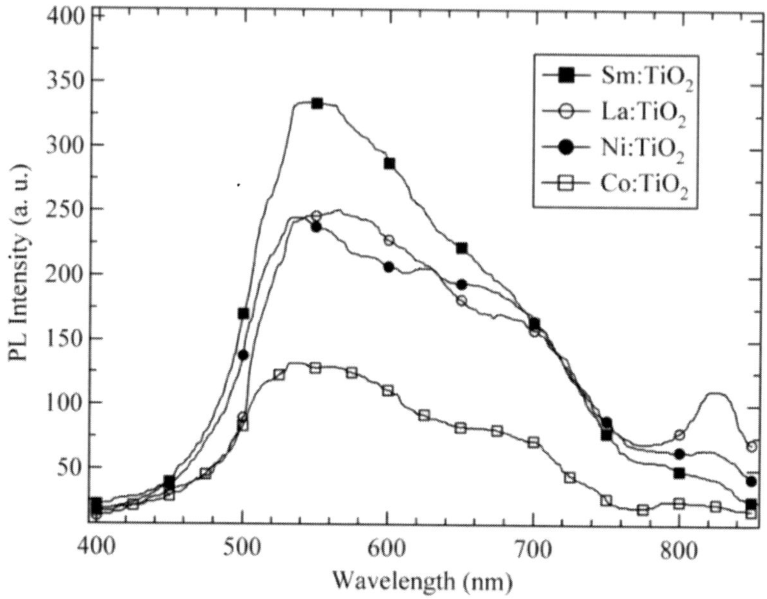

Figure 6. Photoluminescence emission spectra of thin films *M*:TiO_2/quartz, *M* = La, Sm, Ni, Co.

Photoluminescence emission spectra of thin films *M*:TiO_2/quartz, *M* = La, Sm, Co, NI are shown in Figure 6. Shapes of the spectra are quite similar, but they differ in the intensity of the bands.

The luminescence band of thin films *M*:TiO_2 is wide and it's located in the region from 420 nm to 850 nm with maximum at 550 nm. Samarium doped titanium dioxide film has the most high intensity and complementary luminescence band with maximum at 830 nm.

Conclusion

The thermal oxidation of two-layers structures M/TiO_2/Si(quartz) results in the compound oxide films formation. Doping metal determines the shift of short-wave transmission band edge and the optical transition energy. La and Sm doping shifts this edge to short-wave region and the edge wavelength equals 330 and 270 nm respectively. Ni and Co doping shifts this edge to long-wave region and the edge wavelength equals 355 and 370 nm respectively. The optical transition energy is equal to 3.70 and 3.06 eV and correspond to $La_2(TiO_3)_3$ and La_2TiO_5 respectively, 4.7 eV for Sm_2TiO_5, 3.64 eV for $NiTiO_3$

and 3.35 eV for $CoTiO_3$. The luminescence band of thin films $M{:}TiO_2$ is located in the region from 420 nm to 850 nm with maximum at 550 nm. Samarium doped titanium dioxide film has the most high intensity and complementary luminescence band with maximum at 830 nm.

ACKNOWLEDGMENT

This work supported by grant-in-aid under federal target program of Russian Federation: "Researches and elaborations in priority directions of development of a scientifically-technological complex of Russia for 2007-2012", project No. 16.513.11.3008.

REFERENCES

[1] G.T.Crotty, P.J.Verlinden, M.Cudzinovic et al., 26th IEEE Photovoltaic Specialists Conference, Anaheim, CA, September 30 October 8, 1997, p. 1035.

[2] Y.Tonomura, H.Washio, K.Nakamura et al., 2nd World Conference and Exhibition on Photovoltaic Solar Energy Conversion, Vienna, Austria, July 6 10, 1998, p. 3511.

[3] H.K. Pulker, *Applied Optics,* 1969–1975 (1979) 12.

[4] Babuji, S. Radhakzishna, *J. Maiter. Sci Lett.* 767–769 (1985) 6.

[5] M. Graetzel, *Comments Inorg. Chem.* 12 (1991) 93.

[6] H. Takikawa, T. Matsui, T. Sakakibara, A. Bendavid, P.J. Martin, *Thin Solid Films* 145–151 (1999) 348.

[7] D.W. Hwang, H.G. Kim, J.S. Lee, J. Kim, W Li, S.H. Oh, *J. Phys. Chem.* B 2093-2102 (2005) 109.

[8] D.L. Wood, J. Tauc, *Phys. Rev.* B 3144 (1972) 8.

In: Smart Nanoobjects
Editor: Kirill Levine

ISBN: 978-1-62808-741-3

Chapter 4

STUDYING CORROSION PROTECTION WITH NANOSTRUCTURED POLYIMIDE COATINGS BY ELECTROCHEMICAL IMPEDANCE SPECTROSCOPY

***Kirill L. Levine*[1*], *Van Fen*, *Natalia N. Nikonorova*[2], *Valentin M. Svetlichnyi*[2], *Vladimir E. Yudin*[2], *Ludmila A. Myagkova*[2] and *Nikolay S. Pshchelko*[1]**

[1]The St. Petersburg State Mining University.
Department of General and Technical Physics, St. Petersburg, Russia
[2]Institute of Macromolecular Compounds Russian Academy of Sciences, St. Petersburg, Russia

ABSTRACT

In this study the anti-corrosion performance of polyimides modified with benzoquinone fragment was evaluated and compared with polyimide missing this fragment. The electrochemical impedance spectroscopy was employed to investigate polyimides pyropolymellitimide-oxydianiline (PMDA-ODA), polyimide-benzoquinone (PIQ), and polyamidoimide (PAI). The nano-effect discovered in earlier work was successfully confirmed: the impedance of nanostructured polyimides increased after

* Author to whom correspondence should be addressed. Email: levinkl@hotmail.com

the treatment with the corrosive solution before coatings failure. Therefore nanostructured systems can be utilized for “smart” corrosion protection.

Keywords: Corrosion protection, polyimide, electrochemical impedance spectroscopy, nano-effect

INTRODUCTION

The development of nanotechnologies gives possibility to apply new principles to corrosion protection. One can think about a coating that responds to corrosive environment in “smart” manner: changes its properties under the influence of corrosive media. In this paper we observed an example of such a behavior: an attempt of a coating to recover its resistance to ions after some preliminary failure caused by the permeation of aggressive ions into the coating.

Polyimides are used largely for microelectronics because of their excellent heat withstanding properties, chemical stability, and electrical resistance belonging to the range of excellent dielectrics [1].

These polymers appear to be attractive candidates for corrosion protection because of their good mechanical properties and high barrier properties [2]. Unfortunately, the majority of polyimides possess a strong tendency to absorb moisture that results in substrate corrosion [3]. In order to improve adhesion of polyimides to iron were suggested PIs with quinone functionalities [4, 5, 6] that have shown enhanced corrosion protection behavior. Introducing benzoquinone fragments into polyimide chain also improves thermal stability of this corrosion resistive coatings.

They are able to withstand short term heating up to 550°C (in nitrogen) compared to 290°C reported for “conventional” PI’s [6]. Also the presence of benzoqiunone fragments improves adhesion to substrate.

The purpose of this study was to compare corrosion protection properties of different polyimides using method of Electrochemical Impedance Spectroscopy (EIS). Polymers selected for this study were: pyropolymellitimide-oxydianiline (PMDA-ODA), polyimide-benzoquinone (PIQ), and polyamidoimide (PAI) (Figure 1).

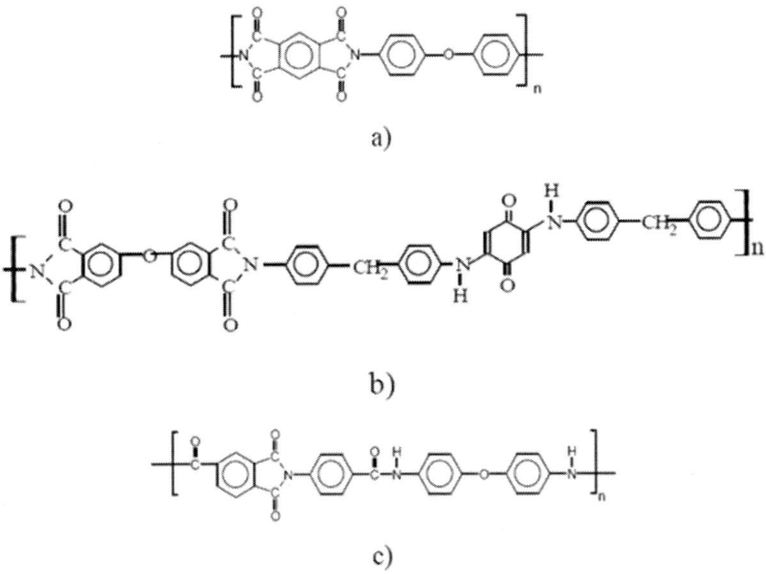

Figure 1. Chemical structures of polyimides used for corrosion protection. 1) Polyimide (PMDA-ODA); 2) Polyimide with quinone fragment (PIQ); 3) Polyamidoimide (PAI).

METHODS

Samples Preparation

PMDA-ODA (pyropolymellitimide oxydianiline) was obtained commercially under the trademark PM in the form of pre-polymer: poly(amic acid) (PAAc). Pre-polymer for PIQ was based on diamine AQMDA ($X=CH_2$); dianhydride ODPA (Y=O), (Figure 2). PIQ and PAI were synthesized at IMC RAS. PIQ was synthesized by a two-step polymerization of amine–quinone monomer (AQMDA) and aromatic dianhydride (ODPA) in 1-Methyl-2-pyrrolidinone (NMP), as shown in Figure 2. Similar procedure was applied to obtain PAI.

Thickness of polyimide films obtained by this procedure was 9-10 μm. All samples were heated at 300^0C to remove solvent and to complete the imidization.

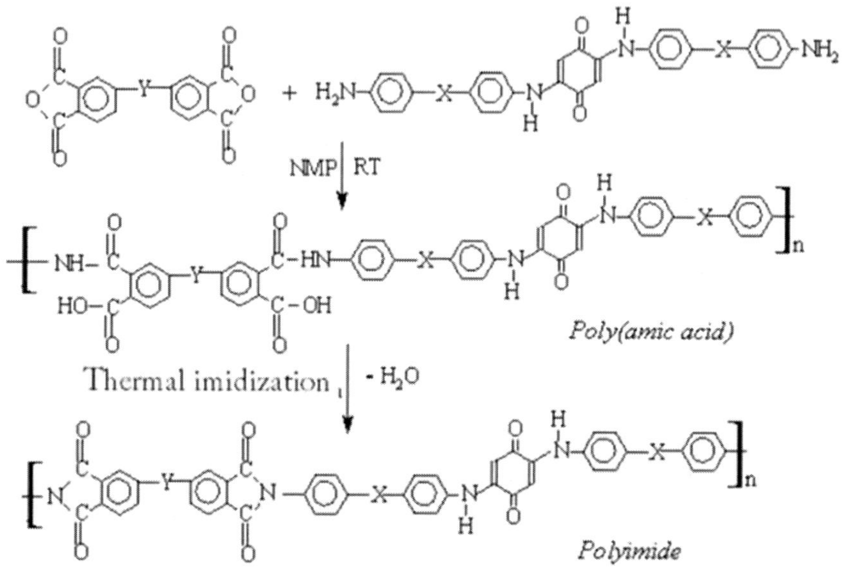

Figure 2. Two-stage synthesis of amino-quinone polyimide PIQ based on diamine AQMDA (X=CH_2) and dianhydride ODPA (Y=O).

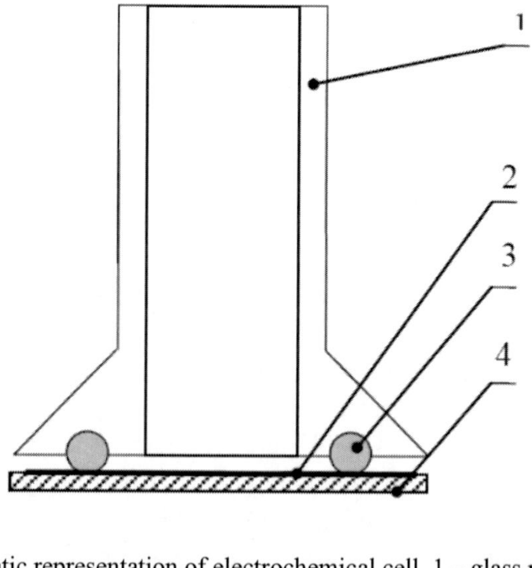

Figure 3. Schematic representation of electrochemical cell. 1 – glass part of a cell, 2 – sample, 3 – rubber ring, 4 – metal plate.

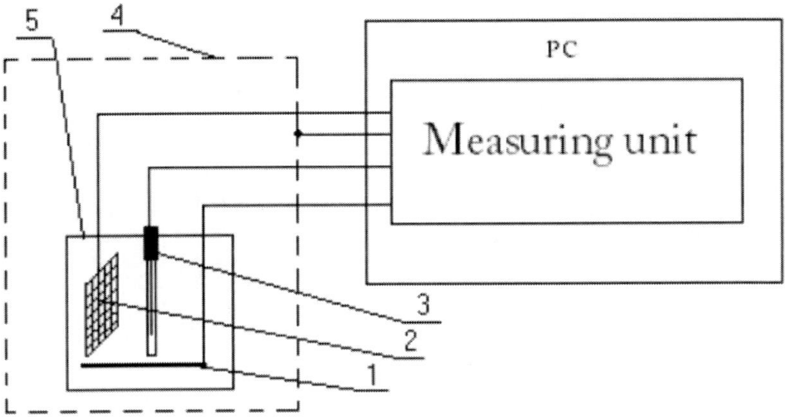

Figure 4. Apparatus used for the experiments. 1 Working electrode, 2 counter electrode, 3 reference electrode, 4 Faraday cage, 5 cell body.

Figure 3 shows schematic of electrochemical cell used in this experiment. Cylindrical cell was attached to a sample with mechanical clamp. It had 3 electrode configuration consisting of working electrode (film coated sample), Ag/AgCl reference electrode and counter electrode made from platinated grid. The cell was protected from electromagnetic interferences by Faraday cage. Steel panels of rectangular shape coated with different polymers were used for the experiments.

Cell was filled with 3% (mass) aqueous solution of sodium chloride. Electrochemical impedance spectra were recorded after different time of immersion of the sample into corrosive electrolyte. Experiment lasted 10 days. Measurements were taken in frequency range from 1 mHz to 5 kHz.

Potentiostat PC3 manufactured by Gamry was used for the experiments. Schematic of its connection to electrochemical cell is shown in Figure 4. Amplitude of applying signal at all measurements was 10 mV.

RESULTS AND DISCUSSION

Figure 5 shows the appearance of PMDA-ODA, PIQ, and PAI samples after 167 h of immersion in corrosive media. Figures 6 - 8 show the impedance spectra of different coatings at different immersion times. Figure 9 presents the three-dimensional impedance plot made at 100 mHz for different samples at different immersion times.

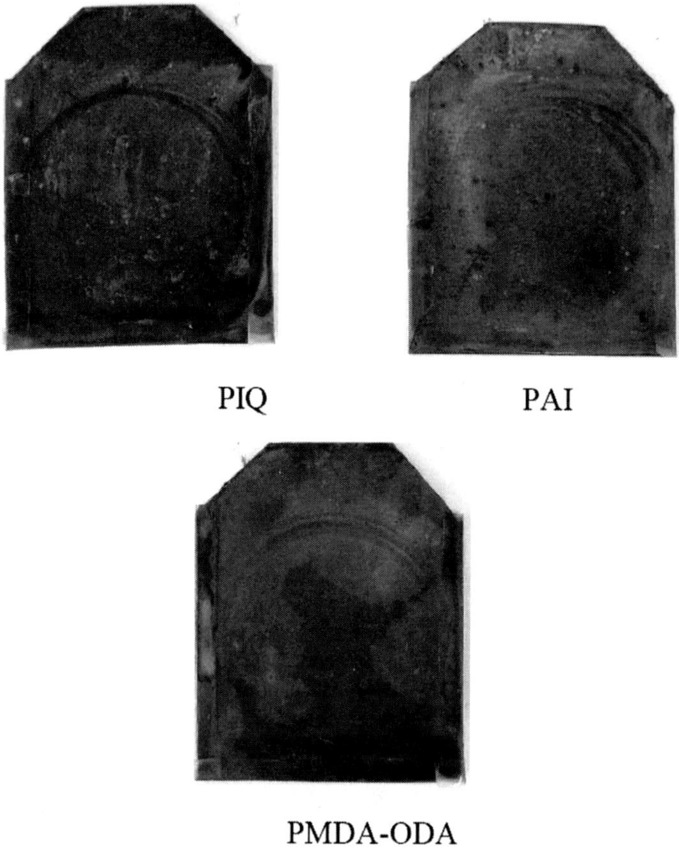

Figure 5. Photographs of samples after 167 h of immersion.

PMDA-ODA shows “classical” behavior (Figure 6). It possessed relatively high impedance immediately after the immersion (>10^5 Ohm). After 215 h it failed by 2 orders of magnitude. Nyquist plot had shown the collapse of semicircle started at 167 h that meant the failure of barrier properties.

Bode plot of PIQ polyimide is shown in Figure 7. Even though the impedance initially possessed larger values than for PMDA-ODA polyimide (10^6 Ohm), it dropped to 10^3 after 167 h. This is worth notation that the impedance from $8.5 \cdot 10^2$ Ohm at 167 h shifted to $1.05 \cdot 10^3$ Ohm at 215 h (by (23±5)%) showing “non-classical” behavior.

Significant non-classical behavior was observed for PAI polyimide (Figure 8). Initial value of impedance of $1.1 \cdot 10^5$ Ohm dropped significantly to $6 \cdot 10^3$ Ohm at 167 h of immersion. However at 215 h it regained to $9 \cdot 10^3$ Ohm that is (50±10)% of the lowest value. This is worth of noting that in the case of PIQ polyimide the non-classical trend was observed within 0.1-10 Hz frequency range while for PAI polymer the trend was observed within the entire frequency range: from 0.1Hz to 5KHz.

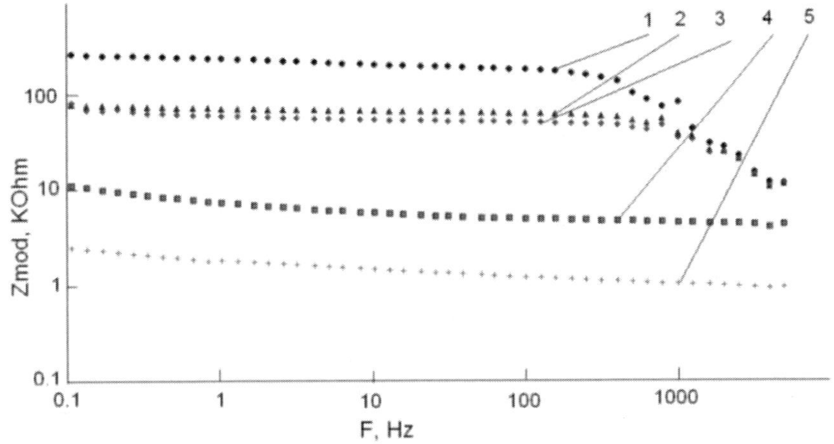

(a)

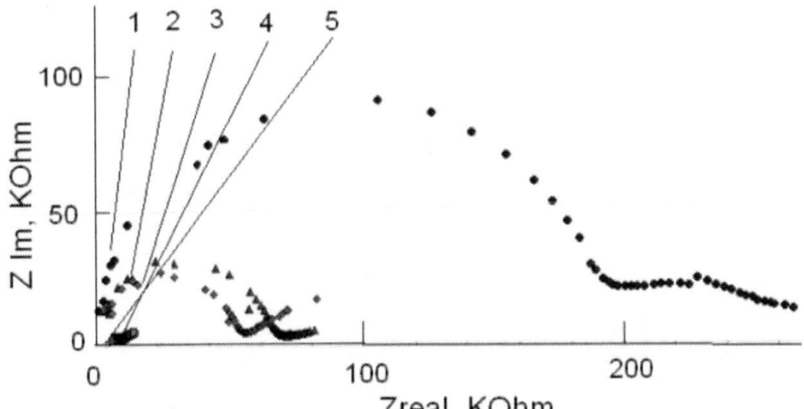

(b)

Figure 6. Electrochemical impedance spectroscopy of PMDA-ODA polyimide. (a) Bode plot, (b) Nyquist plot. 1 – 1 hour, 2 – 22 hours, 3 - 73 hours, 4 – 167 hours, 5 – 215 hours.

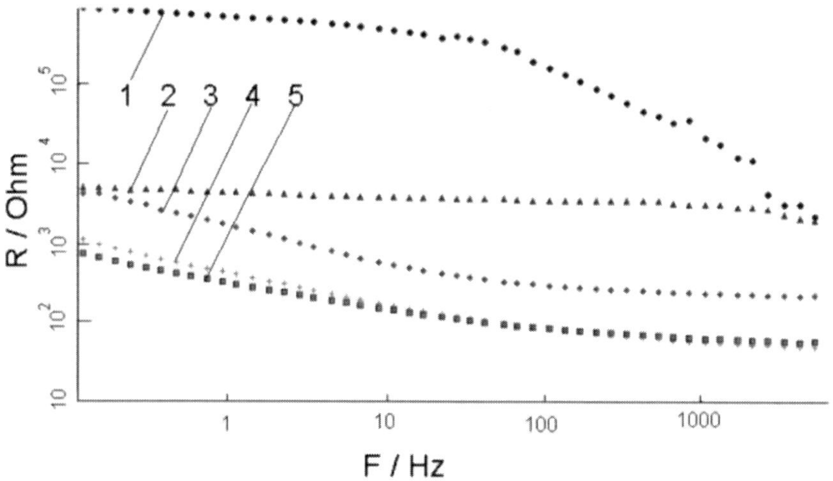

Figure 7. Bode plot of PIQ polyimide. 1 – 1 hour, 2 – 22 hours, 3 – 73 hours, 4 – 215 hours, 5 – 167 hours.

3D plot of a coatings modulus is shown in Figure 9 reveals that PIQ possess the poorest corrosion protection properties while PAI and PMDA-ODA the best ones. At long-lasting forecast PIQ looks more promising than PMDA-ODA.

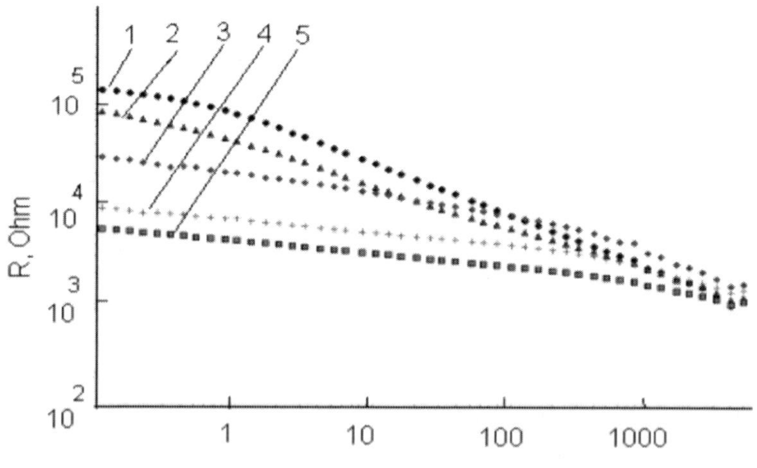

(a)

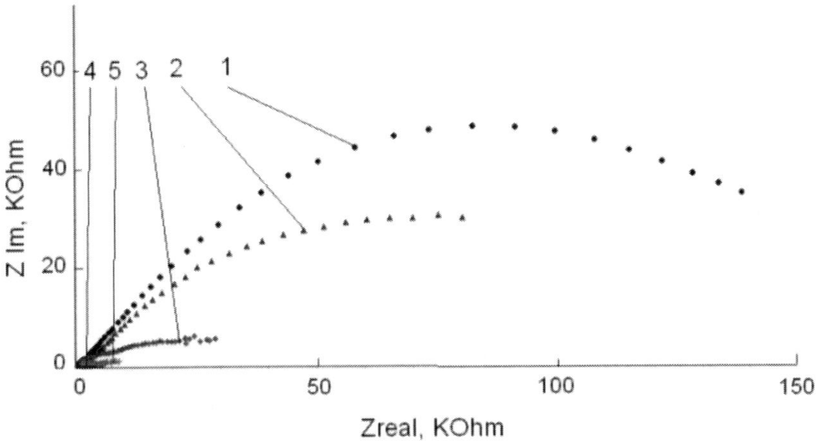

(b)

Figure 8. Electrochemical impedance spectroscopy of PAI polyimide. (a) Bode plot, (b) Nyquist plot. 1 – 1 hour, 2 – 22 hours, 3 - 73 hours, 4 – 215 hours, 5 – 167 hours.

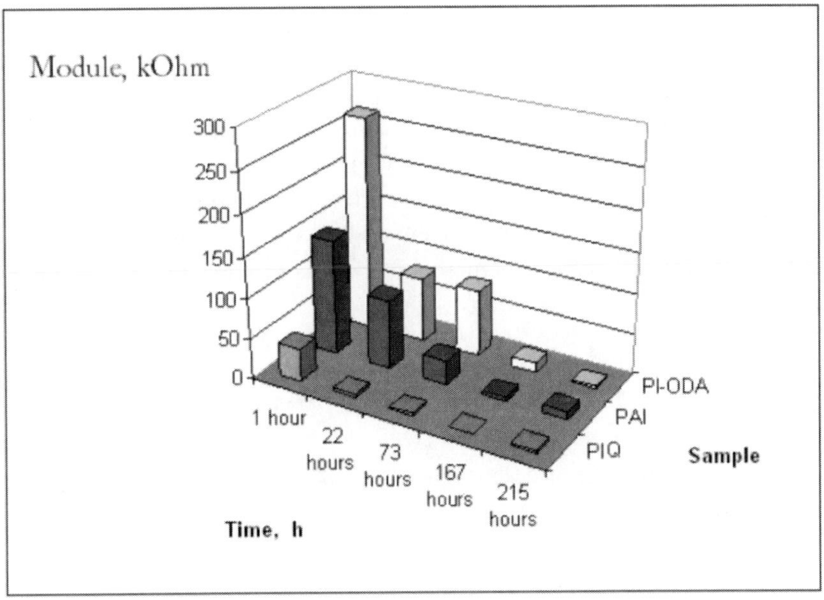

Figure 9. Impedance data for different samples at 100 mHz at different immersion times.

This interesting result can be attributed to a different type of response of PIQ polyimide to the ingress of aggressive ions into the coating and its swelling. Similar effect was reported in [7]. This effect was not observed for “classical” polyimide PMDA-ODA even though experiments were experiments were repeated few times. For PAI and PIQ the trend was repeatable. In PIQ and PAI the non-classical behavior possibly arises from the ability of molecular clusters to absorb moisture and to interrupt the diffusion through the mechanism of changed pore resistance [8] as a result of changing pore size [9].

Smart behavior of some polyimides is possibly related to the ability of their nanostructured morphology to adjust permeability to aggressive solution dynamically: change porosity as a result of swelling of nanostructured aggregates. More detailed close-look at the described effect will be performed in further investigation.

CONCLUSION

The comparison of PAI, PIQ and PMDA-ODA polyimides have revealed that these 3 different polymers possess different corrosion protective properties, show different response to corrosion. It was determined that PIQ and PAI polyimides show “nano-effect”. After certain time of immersion their impedance increases quite significantly possibly due to the swelling of nanostructures under the ingress of corrosive liquid. Therefore, further investigation of PAI and PIQ with nano-additives will be promising from the point of view of using nanostructured materials as smart coatings for corrosion protection.

ACKNOWLEDGMENTS

Acknowledgements are brought to colleges from the Institute of Macromolecular Compounds Russian Academy of Science: Vladimir E. Yudin, Valentin M. Svetlichnyi, and Ludmila A. Myagkova for preparation of nanostructured samples.

REFERENCES

[1] Bessonov, M.M.Koton, V.V. Kudryavtsev, L.A. Laius. Polyimides-Thermally Stable Polymers, *Plenum Publishing Corp.*, New York, 1987.

[2] Sazanov Yu. N. Applied Significance of Polyimides, *Russian Journal of Applied Chemistry*, 2001, v. 74, No 8, p. 1266.

[3] Tan, E. K.; Bellucci, F.; Kloppers, M. J.; Latanision, R. M., *Mater Sci Forum* 1992, 111–112, 177–190.

[4] Kaleem, K.; Chertok, F.; Erhan, S. *Prog. Organ. Coat* 1987, 15, 63–71.

[5] Nithianandam, V. S.; Chertok, F.; Erhan, S., *J. Coat. Technol* 1991, 63, 47–49.

[6] Han N. M., Bie H., Nikles D.E.,Warren G.W., *Corrosion Journal of Polymer Science: Part A: Polymer Chemistry,* Vol. 38, 2893–2899 (2000).

[7] Tallman D.E., Levine K.L,Siripirom Ch, Gelling V.J., Bierwagen J.P., and Croll S.G., *Applied Surface Science*, 254 (2008) 5452 – 5459.

[8] Levine K.L. and Iroh J.O. "Resistance of the Polypyrrole/Polyimide composite by electrochemical impedance spectroscopy", *Journal of Porous Materials*, 11, (2004), 87-95.

[9] Levine K.L., Pshchelko N.S., Electrochemical Properties of a Polypyrrole-Polyimide Composite, *Polymer Science series A (Polymer Physics),* 2011, Vol. 53, N 6, p 510-520.

In: Smart Nanoobjects
Editor: Kirill Levine

ISBN: 978-1-62808-741-3

Chapter 5

SURFACE-NANOSTRUCTURED METALS AND THEIR TRIBOCHEMICAL PROPERTIES

*A. G. Syrkov**
Saint-Petersburg State Mining University, St. Petersburg, Russia
Scientific-educative Center of Nanotechnologies, St. Petersburg, Russia

ABSTRACT

The possibilities of the adsorption modification of the surface of metal powders in vapors of quaternary ammonium compounds for regulation of the tribochemical properties of metals and lubricants containing these metals as additives was analyzed. The general trend is that in order to achieve a minimal index of friction (D) of tribological pair with a lubricant, a combination of few properties of compounds is necessary. One is good water-repellent, and another is low reactivity. It was revealed that solid-state synthesis of metals in vapor of hydrophobic silicone liquid based on organohydridesiloxanes is perspective for obtaining Si-C-containing surface-nanostructured metallic composites (M = Ni, Cu, Fe) with high temperature stability. The mechanism of this synthesis, according to XPS and other techniques, allows to regulate with precision the structural and chemical characteristics of metallic materials, including values of electron energy in the Si2p surface layer level in the range 102-106 eV. Maximum of high temperature stability in air has a sample on the basis of iron.

* E-mail: syrkovandrey@mail.ru

Keywords: Adsorption modification, solid-state hydride synthesis of metals, metal powders, nanostructured metallic composites

INTRODUCTION

The problems of stability properties of metals and improvement the quality of metals are important not only in their preparation, but also in the process of storage and functional use. The urgency of these challenges in the XXI century is increased in connection with the demand of industry in dispersed and nanostructured metallic materials. Disadvantage of latter, as a rule, is low stability in air and in aggressive technogenic environments. Metals as thin film and a highly dispersed belong to the class of metal-dielectric structures and metal-dielectric-semiconductors (MDS), which are the basis for many intelligent electronic devices. In addition, nanostructured metal powders are used for filling and properties improvement of various organopolymer compositions [1, 2].

Most of being developed today nanotechnology methods of metals are often quite expensive and energy consuming (electrical explosion, synthesis in plasma, plastic deformation of metals). The main thing - they do not solve the central problem of modern solid-state production: to provide reliable reproducibility of structure and properties of materials from one sample to another and from synthesis to synthesis.

In some cases quite successful and proven can be approach that is based on the irreversible chemisorption in terms of various substances on the surface of dispersed or compact metals [2, 3]. In this case, it will be more accurate to talk about getting not just a nanostructured materials (nanocomposites) but the synthesis of surface-nanostructured metals [2]. In the most general case, the metal-substrate does not need to be a nanostructured material, and may represent a roughly dispersed metal powder or, for example, conventional steel [4, 5]. Many of metals applications (adsorption, catalysis, lubrication, friction and electronics) impose sufficiently high requirements to the surface material properties including its thermal and chemical stability. Therefore, it seems logical to apply nanotechnology methods that will lead primarily to improved quality of metal surface.

In this paper some particular qualities of tribochemical properties of surface-nanostructured metals (Al, Cu, Ni, Fe), obtained by solid-state hydride synthesis and the adsorption modification of metals by cationic tensides based on quaternary ammonium compounds are considered. The above methods of synthesis and surface modification of metals were developed under the guidance of the author [2, 4-8, 9].

Solid-state hydride synthesis of metal materials is a reduction in an open flow system under the specified program of solid metal compounds by thermally stable volatile elementohydride compounds (NH_3, CH_4, SiH_4, CH_3SiHCl_2, etc.). This method is based on laid out solid metal halides and oxides (Ni, Fe, Cu, Zn) reduction by heating. By this method we have produced metals and new surface-nanostructured dispersed metallic materials [2, 4, 6]. Conditions of synthesis guarantee that the adsorption of reducing agent on metal do not exceed the value of monolayer adsorption. From the measured XP-spectra of samples it follows that adsorbed elementohydride chemically interacts with the reduced metal, i.e. reductant's molecules are chemisorbed [2, 4]. The peculiarity of our proposed method is that depending on the selection of a reducing agent, direct control of the specific surface of solid material, its thermal and chemical stability, electron binding energy of the characteristic level of surface atoms (XPS data) is possible [4 ,6]. Of particular interest is the reduction of metals in vapor of methyldichlorosilane (CH_3SiHCl_2) or organohydridesiloxanes with subsequent heat treatment in a flow of methane (natural gas). As a result of this synthesis metallocomposites are formed with Si-C-containing protective surface nanofilm, which gives metal a unique heat resistance, good water-repellent and organophilic properties [2, 8].

I. Surface-Nanostructured Metals Obtained by Method of Adsorption Modification and Their Properties

For surface treatment of metals from aqueous solution and vapor phase we used alcamon (A) and triamon (T) as a substance-modifiers, which in Russia are traditionally used to impart antistatic properties to the textile and polymer materials. The composition of T corresponds to the chemical formula $[(HOC_2H_4)_3N^+CH_3][CH_3SO_3^-]$. Alkyl radical in alcamon's cation corresponds to the number of carbon atoms n = 10 ÷ 18. The original aluminum powder (specific surface 2,6 m^2/g), copper (0,6 m^2/g) and carbonyl nickel (0,4 m^2/g) were treated in A and T vapor [2]. Metals were treated with either one preparation (A or T) or a mixture of (A + T) or sequential interaction of metal (M) with T and A. We have obtained a series of samples, respectively, of the form: M/A, M/T, M/(A + T), M/T/A, where M = Al, Ni, Cu. In addition, as a sample of comparison we used the powder form M/HSL. This powder is dispersed metal treated by vapor of hydrophobic silicone-organic liquid (HSL). The magnitude of adsorption of water vapor (aH_2O) was measured by the dessicator method under P/P_0 = 0,96-0,98 (20 ± 2^0C). Reactivity determined gravimetrically by measuring specific weight gain during the oxidation of samples in a furnace (1113 K, 300s) in the free access of air (P = 1 atm, humidity is 60 ± 10%). The antifriction properties of powders were evaluated by acoustic emission method. Integral index D (proportional to the friction force) in the tribological pair drill - a steel plate with a lubricant that contains a metal powder was studied [2]. D was measured on a special stand in the ultrasonic range of frequencies (20-300 kHz) with the help of a certified device ARP-11. The content of the powder additives in the lubricant was 0,5-1,0 mass%.

The analysis of three-dimensional relationships between integral friction index D and reactivity and the magnitude of adsorption of water vapor is shown in Figures (1-3) and Tables (1-3). Dependencies are 3dimentions space curves that reflect the influence of surface modification of metal on the antifriction properties of lubricant under boundary friction. The simplest dependence we have for the lubricant filled with Al- or Ni-samples. There is a general trend: D decreases (friction decreases) with decreasing reactivity of the filling powders (Figures 1-4). In the case of powders based on aluminum it is observed that D decreases, as a rule, with decreasing values of aH_2O (i.e. with increasing hydrophobicity of the samples), as shown in Table 1.

Table 1. Values of and aH_2O for Al-samples

Filling powder	aH_2O, %	D for industrial oil (I-20) with powder (p=47 MPa)
Al/T	0.17	780
Al/A	0.22	910
Al/ (A+T)	0.13	300
Al	0.23	1690
Al/T/A	0.16	1000

For convenience of comparison with the data of Figures (1-3) the composition of powder samples in Tables (1-3) is shown by the increase in their reactivity (for example, from Al/T to Al/T/A, Table 1).

From Tables 1 and 2 we can see that the powder M/(A + T) has the stronger effect on friction reduction than the powder form M/A and M/T (M=Al,Ni).

For lubricant with the addition of Al/(A + T) D value is the smallest: 300, for lubricants with additives Al/A and Al/T respectively, D is equal to 910 and D is equal to 780 (Table 1), respectively .

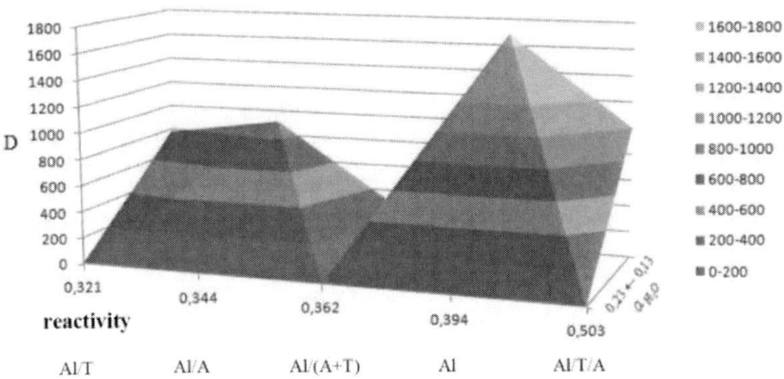

Figure 1. Diagram of D relationship with the value of adsorption of water vapor and the reactivity during the oxidation of the modified metal additive on the basis of Al-powder.

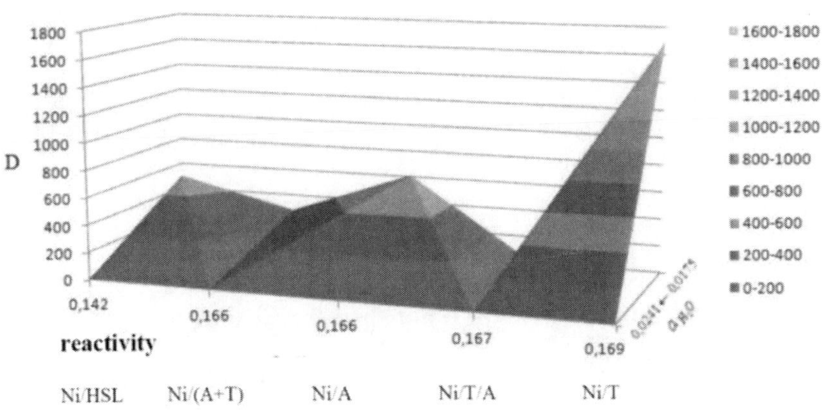

Figure 2. Diagram of D relationship with the magnitude of adsorption of water vapor and the reactivity during the oxidation of the modified metal additive on the basis of carbonyl nickel.

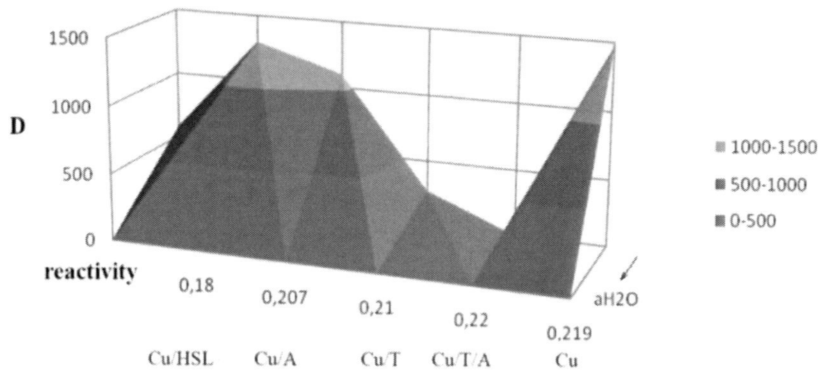

Figure 3. Diagram of D relationship with the magnitude of adsorption of water vapor and the reactivity during the oxidation of the modified metal additive on the basis of copper powder (PM1).

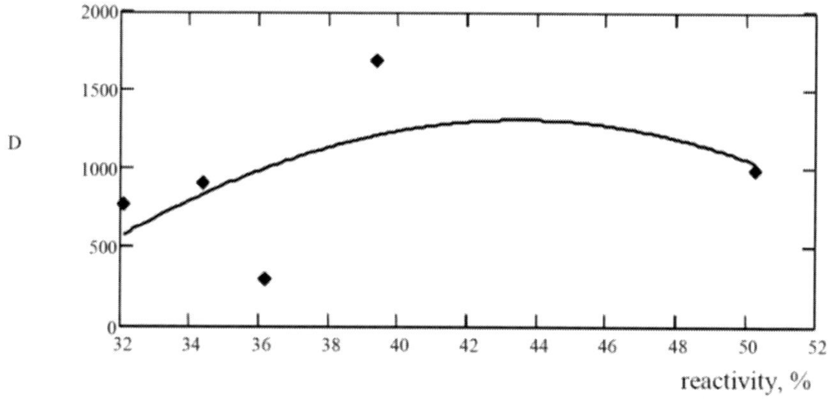

Figure 4. Dependence of D from filler reactivity for lubricant with additives on Al-base. Approximation equation (Math Cad) is $D=7.11y-7.03y^2-4330\ln y$, where y=reactivity.

Table 2. Values of D and a_{H2O} for Ni-samples

Filling powder	a_{H2O}	D for industrial oil with powder
Ni/ HSL	0.0175	500
Ni/ (A+T)	0.0212	280
Ni/ A	0.0212	610
Ni/T/A	0.0205	-
Ni/T	0.0191	1700 (dry friction)
Ni	0.0241	280

In fact the influence of A and T is synergistic. They act in a manner that the activity of their mixture (A + T) to D exceeds the activity of individual components. Dependence of D on the reactivity and aH_2O, as can be seen from Table 3, for Cu-samples is more complicated. The synergistic effect on D was observed in the powder consistently treated by T and A (Cu/T/A). On the sample treated by a mixture of modifiers (A + T), in contrast to the Al/(A + T) and Ni/(A + T), a marked increase in D (Table 3) until 1480 takes place.

Table 3. Values of D and a_{H2O} for Cu-samples

Filling powder	a_{H2O}	D for industrial oil with powder
Cu/HSL	0.0205	580
Cu/A	0.0299	1300
Cu/T	0.0268	1100
Cu/T/A	0.0260	270
Cu	0.0445	-
Cu/(A+T)	0.0310	1480

The friction in the system increases. The general trend, which is observed for all modified metallic additives on the basis of Al, Ni, Cu is the following. To achieve the minimum D (lowest friction) in the tribological pair with a lubricant, combination of good water reppelent properties and low reactivity of the additives is favorable. A clear illustrations of this rule are the properties of additive Al/(A + T) in Table 1, Ni/(A + T) in Table 2 and Cu/T/A in Table 3. The reactivity of these additives as is evident from these Tables and Figures 1 – 4 has a sufficiently low level (Ni/(A + T)), or less than the average level of this properties in a series of modified samples on the basis of the corresponding metal (look samples Al/(A + T) and Cu/T/A).

The synergistic effect of D for the latter sample can be explained in the following way. According to modern concepts, in order to achieve the maximum of antifriction effect for solid surfaces treated with surfactants, it is necessary to combine high hydrophobicity of the surface and good adhesion of deposited films to the original surface [2]. Data of our studies testify that from two of these factors, the latter factor (adhesion) is the most important [9]. Metal processing by two types of surface-active substances (A and T), in general, is favorable to the implementation of both significant factors. T molecule with nitrogen atom that contains small organic radicals, interacts with metal surface. This interaction, according to XPS, is possible due to donor acceptor bond with the transfer of electrons from N to M [2, 4, 9]. Due to this bond, it is obvious that there is a stabilization (increased adhesion) bilayer film T / A on metal surface after consistent treatment with mentioned modifiers, which have structural compliance, as both are quaternary ammonium compounds (QAC). Ensuring good adhesion of the film T / A under similar treatment of aluminum and nickel is difficult because their initial surface contains organic impurities [2]. Used Al-powder (trade-mark PAP-2) has a stearin nanofilm on its pre-coated by the factory. Selected for the experiments carbonyl nickel powder (PNK-UT3) is also characterized by a high content of carbon (carboxyl groups) in its surface layer.

In terms of the boundary and dry friction at high load pressures when the liquid lubricant (oil) is squeezed out of zone of tribological contact drill-plate, the measured friction index D to a large extent is determined by the antifriction properties of solid surfaces metallic additives [2]. This obviously explains the fact that the minimum of D for lubricant with copper additives is observed on the powder type Cu / T / A (Table 3).

When using additives based on Al-powder the lubricant with sample Al / (A + T) has the lowest D . The advantage of processing additive mixtures, is apparently due to the fact that the heteroatomic interaction of triamon and metal according to scheme N → M is hindered because of stearin films on aluminum. Triamon-alcamon film can better stabilize the surface of the additive due to structural similarity of hydrocarbon radicals of alcamon and stearin [2, 4].

II. Solid-State Synthesis of Si-C-Containing Metallic Powders and Their High Temperature Stability

Reduction vapor of HSL based on organohydridesiloxanes on the first stage was recently used for obtaining surface-nanostructured SiC-containing metal powders [2]. The advantage of HSL over methyldichlorsilane applied earlier [4, 8] is that HSL is less toxic. Treatment on the second stage of the synthesis in methane is still necessary because it allows to reduce metal contents in the volume of solid phase, where penetration of large molecules HSL is difficult enough. After the reduction of solid metal oxides (MO) or chlorides (MCl_2), obtained metal powders have properties and characteristics presented in Table 4.

Table 4. Structure-chemical characteristics of Si-C-containing metal powders

Sample, way of obtaining	Specific surface, m^2/g	Δm for 100 h (900^0C) on air, mcg/cm^2	Atomic ratio M/Si in surface layer (XPS)	E_{Si2p}, eV	Chemical shift for $M2_{p3/2}$-level, eV
$NiCl_2$+HSL+CH_4	10	0.398	0	104.7	2.5
NiO+HSL+CH_4	61	0.395	0	105.6	1.4
CuO+HSL+CH_4	13	0.301	0.9	103.8	2.6
FeO+HSL+CH_4	2	0.102	1.1	102.5	5.0

Mass increase (Δm) of metal powders during oxidation is at the level of 0,1-0,4 mcg/cm^2, that is in 17 - 70 times lower than Δm for hot-resistant nikel-chrom alloys (about 7 mcg/cm^2). The samples have high hot stability, which is regulated depending on the type of metal. Synthesis also allows to adjust the surface area of powder and binding energy of electrons (oxidation) of the surface atoms of Si (Table 4). The sample most chemically stable during high temperature oxidation on Fe base possessed the lowest binding energy of Si2p characteristic level (102 eV) for silicon in the Si-C-groups of the modifying metal, and the maximum level of the chemical shift M2p3/2 of the relevant metal. According to our ideas [2, 4, 8], it can be interpreted as follows. Metallic iron, as the most chemically active metal, compared with Ni and Cu, most willingly gives up its electrons to the vacant 2p orbitals of silicon. Thus there is a sufficiently strong heteroatomic M → Si interaction between the metal and the protective Si-C-containing (carbosiloxane) film. Iron, in fact, passivated. The corresponding sample therefore has a maximum temperature stability (minimum Δm), which is confirmed experimentally (Table 4). The synthesized powders were studied in details by IR-spectroscopy, RFA, XRD. In particular, transmission electron microscopy has shown that the thickness of the protective Si-C-containing film on the metal was less than 5 nm [2, 9].

CONCLUSION

The possibilities of the surface adsorption modification method of metal in vapor of cationic tenside preparation (triamon and alcamon) with very different-sized hydrocarbon radicals on the nitrogen atom are reviewed. There are established routes of strengthening of hydrophobicity, and reactivity of surface-nanostructured metals depending on the program of modifying of powders. A general trend of changing of the integral index of friction D for a tribological pair with a lubricant containing various powders modified metals was identified. To achieve the minimum D (lowest friction) favorable combination of good water repellent surface properties and low reactivity of metal powders is needed. Synergistic effects of antifriction properties of lubricants which were observed when using the Al, Ni, Cu based powders additives were detected and interpreted. Found relationships (D-adsorption of water vapor - the reactivity of the powder) are of interest for the selection of perspective antifriction additives on the basis of ductile metals for industrial oils. Based on these results are simple gravimetric measurements of adsorption properties and reactivity of dispersed metallic additives.

Solid-state hydride metal synthesis in HSL vapors of is promising for the creation of new surface-nanostructured Si-C-containing metal-composites with unique temperature stability. The mechanism of the synthesis, according to XPS and other techniques, allows the fine regulation of the structural and chemical characteristics of the obtained metallic materials based on Ni, Cu, Fe. Maximum temperature stability in air (Δm = 0,1 mcg/cm^2 under 900^{0}C, 100 h) was shown by a sample on the basis of iron.

Acknowledgments

The author would like to thank post-graduate student L. Yachmenova for help in getting up of article materials.

The work was supported by Analytical Department Program «Development of scientific potential of high school» (2009-2011) of Ministry of education and science RF, project № 1.13.08, № state reg. 0120.0852107.

References

[1] S. Thomas, G. Zaikov, S. Calsaraj, A. Meere, Recent Advances in Polymer Nanocomposites: Synthesis and Characterisation, *CPI Antony Rowe*, (2010).

[2] I.N. Beloglazov, A.G. Syrkov, Chemicophysical bases and methods of obtaining of surfacely-nanostructured metals, *St. Petersburg State University*, (2011).

[3] V.B. Aleskovsky, Stehiometry and synthesis of solid compounds, Nauka, (1976).

[4] A.G. Syrkov Methods of Physics and Chemistry in Obtaining of Nanostructured Metals and in Nanotribology, Non-ferrous Metals. *Nano-Structured Metals and Materials*, 4, (2006), 10-15.
[5] I. Pleskunov, A. Syrkov, D. Bystrov, On uniform principles and ways of creation of nanostructured metallic and antifrictional materials on steel base, *CIS Iron and Steel Review*, 1-2, (2008), 23-25.
[6] A. Syrkov, Hydride solid state synthesis of dispersed metal and metal-ceramic materials with chemical stability, *Proc. Int. Symp. on Reactivity of Solids*, (1996), 31-33.
[7] I. Yartsev, I. Pleskunov, A. Syrkov, D. Bystrov, Interrelation of water-repellent and protective properties of coatings on steel and role of nanostructured additives, *CIS Iron and Steel Review*, 1-2, (2008), 26-29.
[8] A. Syrkov, L. Mahova, Hydride synthesis of SiC-containing metallic composite materials, *Proc. of Int. Conf. on Electronic Ceramics – Production and Properties*, (1990), 113-115.
[9] N.S. Pshchelko, A.G. Syrkov, T.G. Vahreneva et al., Electrophysical and chemico-physical micro- and nanotechnology of adhesion's enhance of the components in the metal- dielectric system, *Russian Nanotechnologies*, 11-12, (2009), 42-47.

In: Smart Nanoobjects
Editor: Kirill Levine
ISBN: 978-1-62808-741-3

Chapter 6

MULTICOMPONENT MASS TRANSFER KINETICS IN THE MODEL OF NANOCOMPOSITES WITH BIFUNCTIONAL MATRIX

Anatoliy Kalinitchev *
Institute for Phys. Chem. and ElectroChem.,
Rus. Academia of Science, GSP, Moscow, Russia

ABSTRACT

The aim of this theoretical investigation is to describe the created modern Model for the multicomponent mass transfer kinetics in the Nano-Composites (NC) - novel materials with the bi-functional matrix. The detailed description of the properties of the such type of novel NC materials: Metal–Ion Exchangers including the methods of their synthesis has been published recently in the Russian recent monograph. The created new theoretical Model is assigned for the investigations of the multi-component mass transfer kinetics in the novel bi-functional NC materials. This Model includes into the consideration the key conception: two co-existing ways routes (**I** - chemical reactions onto active NC nano-sites, formed by NanoParticles (NP) agglomeration and **II** - diffusion mass transfer) inside the bi-functional NC matrix. The mass transfer process in the bi-functional NC matrix has been considered theoretically

* E-mail: kalina@phyche.ac.ru.

by computerized simulation. The results of the calculations are new and illustrated by author's animations showing visually the propagation of the multi-component concentration waves inside the NC matrix: **r**-beads, cylindrical **ro**-fibers, or planar **L**-membranes. The results demonstrate rather comprehensive analogy between theories of nonlinear chromatography and multicomponent kinetics in the bi-functional NC matrix.

Keywords: Nano-composites, mass transfer, concentration waves, diffusion, multi-component kinetics, bi-functional IEx matrix, active nano-sites

NOMENCLATURE

Symbol	Meaning	Units
T =	$D_{0*} t / r_0^2$, $D_{0*} t /(ro_0^2)$, $D_{0*} t / L_0^2$ – time (t)	dimensionless
D_0	scale for diffusion coefficients	D_0 , [cm^2/sec]
D_i	Diffusion coefficients (relative to scale - D_0); $D_{Ri} = 0$	dimensionless
r =	r/r_0 - distance r along radius r_0 of the **r**-bead	**r**, dimensionless
ro =	ro/ro_0 – distance ro along radius ro_0 of a cylindrical **ro**-fiber	**ro**, dimensionless
L =	l /L_0 - thickness l of planar **L**-membrane, L_0, cm	**L**, dimensionless
X_i	concentrations of i,j-components (relative to input conc.- C_0)	dimensionless
NC	NanoComposites	
i	Index, number of i- component , i = 1,2,…	
[X_j]	Concentrations of j-comp. (in MAL); n - component NC, IEx systems ; computerized NC simulations	dimensionless
NP	NanoParticles (NP) inside NC matrix	
Me^0 $\mathbf{R}^0$	Me^0 (Metal), zero valent Me^0 nano-sites in NC $_k\mathbf{R}^0$ (k = 5 or 6) – zero valent nano-sites in NC	

Symbol	Meaning	Units
[$_k\mathbf{R}^0$]	concentration of $_k\mathbf{R}^0$ - component wave, k =5 or 6	dimensionless
MAL	Mass Action Law (term , irreversible thermodynamics)	
K_S	constants; "association-dissociation" relations (MAL)	dimensionless
n	n –components in NC; IEx systems for modeling	
n=5	Variant 1, Five components in NC, IEx system with 2 MAL reactions	
K_A	constant for 1 st MAL reaction (in pores), Variant 1	dimensionless
K_B	constant for 2nd MAL reaction (at center-sites $_5\mathbf{R}^0$), Variant 1	dimensionless
n=6	Variant 2, Six components in NC, IEx system with 3 MAL reactions	
K_1	constant for 1st MAL reaction (in pores)	dimensionless
K_2	constant for 2 nd MAL reaction (at center-sites $_6\mathbf{R}^0$)	dimensionless
K_3	Constant for 3^d MAL reaction (at center-sites $_6\mathbf{R}^0$)	dimensionless
[$_1SO_4^{2-}$] [$_1Cl^-$]	concentrations of Anions (i=1, index to the left) for: n=5 (Variant 1) or n=6 (Variants 2) ; NC , IEx n-component system	dimensionless
CM - Integral Parameter for "wave" profile (**CM**$_R$ - points on the T axis (abscissa); the frontal $_k\mathbf{R}^0$ – wave is divided according to "equal areas" rule; **Disp**$_R$ – Integral width of $_k\mathbf{R}^0$ - wave		
CM$_R$(T)	"Center Mass" of frontal $_{5,6}\mathbf{R}^0$ - concentration wave	dimensionless
Disp$_R$(T)	Dispersion (width) of frontal $_{5,6}\mathbf{R}^0$ - concentration wave	dimensionless

1. Introduction/Background

Theoretical aspects for the synthesis of the Nano-Composites (NC) with a number of practical applications have been discussed in details in the Russian recent monograph [1]. The modern theoretical Model was created for the simulation of the multicomponent kinetic process in the bi-functional NC matrix [2]. The NC include a lot of the NanoParticles (NP) embedded in advance inside the matrix-medium of the NC. The details of the process of the NC synthesis with zero valent Metal (Me^0) in the ion exchanger matrix are presented [1]. Synthesized NC with the bi-functional matrixes, as for example: "Me^0 - Ion Exchangers", have a lot of practical advantages in comparison with the practice of the usual IEx materials [1, 3].

The corresponding obvious experimental micrography and pictures for the NC structure with the NP are presented (Figure 1*a*-d) as the illustration of the NP agglomerates embedded into the NC media [1]. Figures 1*a, b* show the experimental microphotography of the structure for the NC "Me^0 - Ion Exchangers" including zero valent $Metal^0$ NP (Figures 1, shaded, 5) inside the NC matrix. The microphotography of the experimentally synthesized NC (Figs. 1*a*, b) is obtained by prof. T. Kravchenko group (Voronezh State University) [1]. The experimental examples are presented as the illustrations of the NC media (Figures 1*a*, *b*). The Me^0 - NP (Figures 1*a*, b, light points; c, d – shaded, 5-Me^0), located near the fixed **R** groups (Figure 1d) of the bi-functional NC matrix play the role of the active nano-centers-sites.

There is proposed the modern multicomponent kinetic theoretical Model created for the NC with the key conception: two co-existing ways for the multicomponent mass transfer inside the bi-functional NC matrix-medium [2]. These two co-ways reflect the two simultaneous processes: (**I**) - sorption-desorption for the components at the active nano-sites of the NC matrix and (**II**) - diffusion of the mixture components in the pores of the NC matrix (Figure 1e). Along with the diffusion mass transfer (**II**) the "sorption-desorption" process is realized in the NC Model via the mechanism of the "association-dissociation" reactions (**I**). Figures 1(c-d) represent schemes-illustrations of the NC bead [1] with all accompanying mass transfer elements. Figure 1e shows the scheme of the two accompanying co-ways: **I** – active nano-sites of the NC with the "sinks" (**I*a***) and "sources" (**I*b***) of the [ij] - mass component "transformation": ij $\leftrightarrow$ i + j (**I*a*,*b***) and **II** - the diffusion mass transfer in the pores of the NC matrix [2].

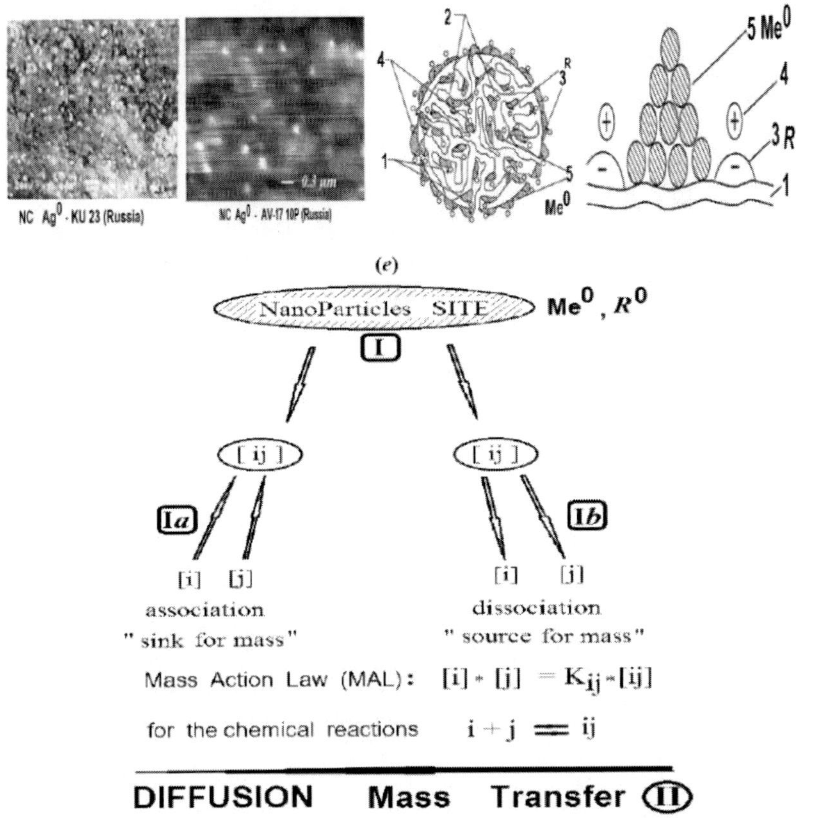

Figure 1(*a*-e). Schemes of mass transfer inside the bi-functional NC matrix : (*a,b*) – experimental NC microphotography, (c) - picture of NC r-bead; (d) - 1-polymeric chain; 2-micro-pores; 3-fixed groups *R*; 4- counter-ions; 5-Me^0-NP agglomerates (c-e, shaded)[1] ; (e) - scheme of mass "transformations" at active nano-sites (Me^0, R^0): (I*a*) – association stage of reactions (left) ; (Ib) – dissociation stage of reactions (right); (II) – diffusion mass transfer along the NC pores.

The various components participate in the reaction (**I**) and diffusion mass transfer (**II**) inside the NC matrix during kinetic process. The mechanism of the "association-dissociation" reactions (**I**) (Figure 1e) brings the retardation of the whole kinetic process. The typical composition of the components may include: ions, complexes, neutral substances with zero charges and, besides the immovable **m**-components (with zero diffusivity, $D_{\mathbf{m}}=0$) at the fixed Me^{0} (or $\boldsymbol{R}^0$) nano-sites (Figure 1e) of the bi-functional NC matrix.

The created modern Model [2] has common points with the preceding approaches published previously at the end of the last century for IEx kinetics

[4-8]. However, the previous kinetic models have been applied to the IEx kinetic processes, accompanied by the chemical reactions only in the usual IEx resins (for **r**-beads), but not for the new NC materials, which are shown in Figures 1(*a-e*).

The modern theoretical Model [2] is used for the computer simulation of the multicomponent mass transfer into the bi-functional NC matrix of the three various shapes: **r**-beads, cylindrical **ro**-fibers or planar **L**-membranes. There is used here the modern NC: Me^0 - Ion Exchanger [1], as the real example for the Model applications. For the generalization of the calculated examples (Section 4, below) the active nano-sites hereinafter are labeled as the k- component: $_k\boldsymbol{R}^0$ (number of k component is left index). So for any NC the $\boldsymbol{R}^0$ label means absolutely the same as Me^0 for the example "Me^0-Ion Exchanger" in the Figures 1(c-e).

The created Model with the key conception (**I,II** - co-ways for mass transfer, Figure 1e)[2] is represented by the adequate computerized description of the mass transfer in the bi-functional NC matrix of the novel NC materials. There are taken into the consideration considered two real factors of the multi-component mass transfer in the NC materials: diffusion ($D_{i,j}$) of the *i,j*-components, together with the coexisting way (**I**) for the "sinks and sources" of the masses (Figure 1e) onto the active nano-sites (Figures 1c-e, dashed) of the bi-functional NC matrix. The specific role of the reaction factor (way **I**) concludes in the retardation of the whole diffusion process due to the equilibrium of the association-dissociation reaction at the nano-sites of the NC (Figure 1e). There is the full similarity between the reaction influence for the retardation of the diffusion kinetics in the NC and the sorption isotherms factor influence in the theory of chromatography [13, 18, 19]. Such similarity will be discussed in details in Sections 6,7. The multicomponent kinetics in the bi-functional NC matrix represent the process, in which the reactions (i.e. equilibrium transformations of various components – factor **I**) must undoubtedly be taken into account for the consideration considered of the mass transport in the NC materials.

The IEx kinetics with the actual processes occurring via the mechanism of "sinks and sources" of masses are discussed in details [4-9]. Herewith, the effect of reactions on the kinetic behavior of the NC system may be crucial: the kinetic process rate may be decreased by one-two orders of magnitude with the decisive dependence of the process on the solution concentration changes. Therefore, distinctive changes in the mechanism of the kinetic mass transport process may occur due to the reaction factor influence. Consideration of the active nano-sites influence in the NC matrix (way **I,** Figure 1e), at which changes of the mass for the components occur, is especially relevant in the case of mass transport processes in the new NC materials [1-3].

2. Modelling of Multicomponent Mass Transfer Kinetics in Bifunctional Nc Matrix

Herewith the general phenomenological approach of the non-equilibrium thermodynamics [10] has been used for the multi-component mass transfer in the NC, IEx matrix [2]. The postulates for the various IEx multicomponent Models for IEx kinetics have been used during long-time period (around 50-60 years) starting from early prof. F. Helfferich book [11] through his Review [12] and with emphasizing STATE-OF–THE-ART-REPORT [7], till 80-90th [4-8]. There is enumerated here (in short) the list of mathematical approaches in the description of IEx kinetics: mass balance partial differential equations, quasi homogeneous medium, isothermal process, electro-neutrality, fundamental Nernst–Planck relationship for fluxes, constant diffusion coefficients (D_i) for i-components. As usual, the influence of the gradient of the electric field is expressed via the sum of the j-concentration gradients by using the absence of the electric current relationships [2, 4-12].

In result the flux of each i- ion is described by its own gradient with addition of the multicomponent superposition of the j-concentrations gradients. Such superposition is called “diffusion potential” in the theory of irreversible thermodynamics (Ch. 4) [10].

For the bi-functional NC matrix the additional mass fluxes are formed due to the transformations of the components, which are expressed by MAL_S relations at the NP nano-sites (Figure 1e, MAL). These sinks and sources of masses are taken into the consideration by the corresponding fluxes of i-components and bring the redistributions of masses for various components participated.

The effects of mass transport for the NC Model [2] by the mechanism “sinks – sources” at the nano-sites (Figures 1d,e) are taken into account in the material balance equations due to the introduction of the additional terms describing decrease of a mass (with negative term for “sink”) or increase of a mass (with positive term for “source”).

For further advance of the Model the equilibrium of chemical reactions has been taken into account by using the classical MAL_S relationships (Ch. 2)[10]. In the computerized Model they are presented in the generalized form where expression Π_i (i≠j) is the product of concentrations $[X_i]^{ms}$ (i= 1,2,..); *ns*, *ms* (negative, or positive) - are stoichiometric coefficients of the chemical s-reaction; s - is the corresponding index; K_S - are equilibrium constants in the MAL_S relationships (1), (Figure 1e). In principle the stoichiometric coefficients (*ms*, *ns*) for the chemical reactions equilibrium (1) may be fractional [6]. During the computerized simulations all the *ms*, *ns* values are assumed to be: (±1). However in need all these values may easily be used in the Model as fractional. The modern approach with all mentioned postulates and equations of the Model [2] has been realized by computer modeling. The set of corresponding computer FORTRAN programs has been composed for the simulation of various multi-component IEx mass transfer in the NC systems, describing kinetic behavior inside the bi-functional NC matrix.

$$\Pi_i\{[X_j]^{ms}\,[X_i]^{ns}\} = K_S,\ s = 1, 2, \ldots\ (MAL_S)\,, \qquad (1)$$

The various shapes of the NC medium: **r**-beads, cylindrical **ro**-fibers, or planar **L**-membranes are included into the theoretical computer simulation. Thus there are realized the computerized investigations of the influence of the equilibrium reactions parameters in cooperation with the diffusion coefficients (D_i) of the *i*-components on the kinetics inside the bi-functional NC matrix of the various three types NC shapes mentioned (**r**-beads, cylindrical **ro**-fibers, or planar **L**-membranes). The basic properties in the development of the created generalized Model are provided by accounting for the multicomponent character of the system and by introduction of the new properties of the bi-functional NC matrix based on the proposed two ways key concept (**I** and **II**) including "association–dissociation" mechanism onto nano-sites $_k\boldsymbol{R}^0$ or $\mathbf{Me}^0$ (Figure. 1a-e). For the further consideration let us simplify the generalized approach (1) to be closer to the multicomponent examples, which have been realized here by computers. The mathematical realization of the diffusion and relations for the chemical reactions inside the bi-functional NC matrix is based on the approach with the application of the MAL (1), (Figure 1e) in the NC Model for the arbitrary ***i, j, ij*** - components concentrations: [*i*], [*j*], [*ij*]. The corresponding simple scheme, as for example for the simple reactions: $\boldsymbol{i} + \boldsymbol{j} \leftrightarrow \boldsymbol{ij}$ may be represented by the simple MAL_S relations (for monovalent components): $[i]*[j] = K_s[ij]$ (Figure 1e). The specific case is realized for the p-components, which are transformed into the complex: $_j(R^0p)$ with zero diffusivity D_j=0.

Meanwhile such p – components may participate in the association-dissociation transformation $_i p \rightarrow {_j(R^0p)}$: $_i p + {_k\boldsymbol{R}^0} \leftrightarrow {_j(R^0p)}$ with the following change of the masses. It will be shown (Section 4) that due to this components transformation at the nano-sites ($_kR^0$, Figure 1e), the corresponding $_kR^0$ -component concentration wave (Figures 3-5, solids, curves 5) propagates in the bi-functional NC matrix. The obtained results of the computerized simulation on the basis of the created Model [2] are presented via the well known "multicomponent concentration waves" concept [13-20]. Concentration waves arise and then propagate along the distance (**r**-, **ro**-radius, or **L**-thickness) inside the matrix of the NC during multicomponent mass transfer. The multicomponent waves with their propagation in the bi-functional NC matrix play the decisive role in the description of the multicomponent NC kinetics of the mass transfer. Therefore this well known and widely used "wave" approach [13-25] is shortly reviewed in the next Section 3.

3. Concept of Concentration Waves for Multicomponent Kinetics and Dynamics of Mass Transfer

During the mass transfer kinetics the concentration profiles -"waves" of the components are formed inside the porous media. Figures 2-5 present examples of the waves propagating during the multicomponent NC kinetics. The key concept of "multicomponent waves" is widely used in theoretical description for many scientific fields of the multicomponent transport for various kinetic and dynamic systems. The "multicomponent waves" concept has wide area for applications in such research fields, as percolation processes [17, 18], mechanics of liquids, gas dynamics [21], theory of burning and even street traffic [18, 22, 23] The term "wave" has been used in all these publications [13-25] including the mentioned excellent monograph [13] and the book [23] concerned with car traffic and kinematic waves.

There are phenomenological concepts potentially common to all filtration processes, which can also be extended to a whole series of migration phenomena such as chromatography, sedimentation, electrophoresis and some others [13-25].

The German-Russian review [20] and presentations [24, 25] published in cooperation include the application of the multicomponent wave concept with the description of the **SCT** (**S**urface **C**omplexation **T**heory) Model. The SCT Model for the multicomponent IEx equilibrium had been elaborated in prof. W. Hoell group (Karlsruhe Research Center) at the end of the last century [20].

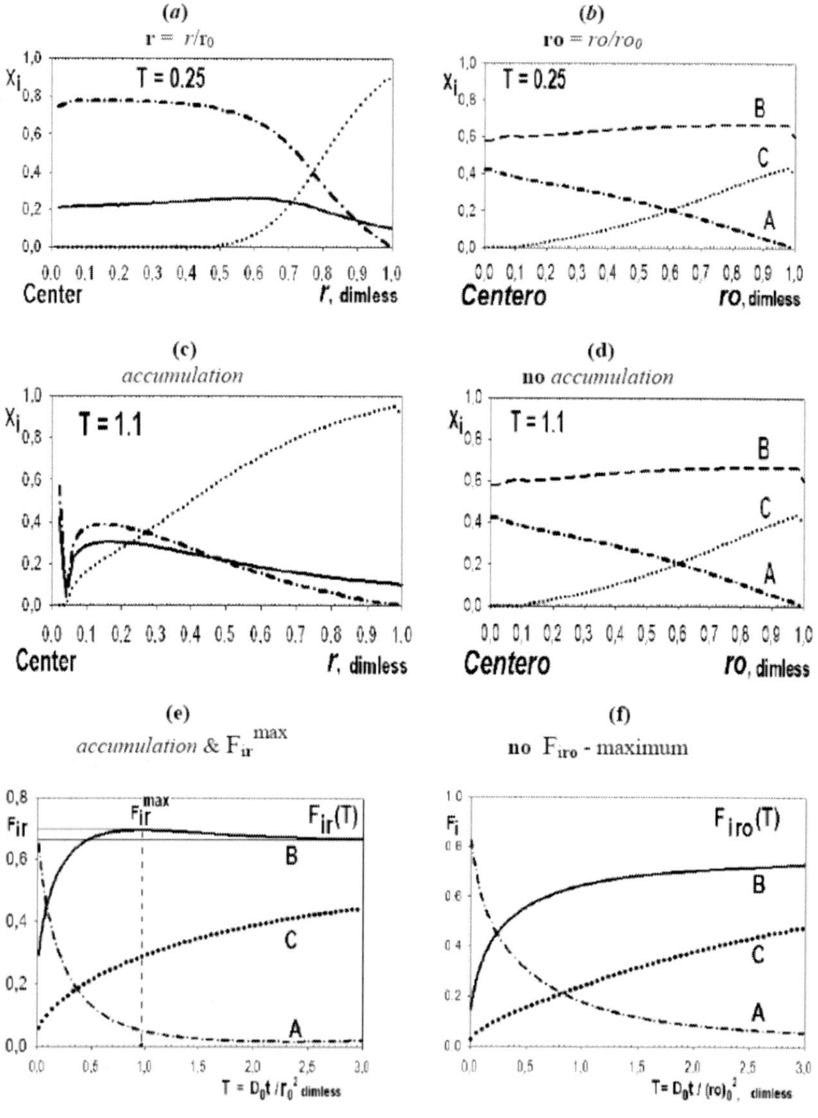

Figure 2(*a*-e). Propagation of concentration waves (*a*-d): X_{ir} (r,T), (left) ; X_{iro}(ro,T), (right) for three ion-components, i = A, B, C. Profiles (*a*-d): A^+- ion (dashed-dotted), B-ion (solid), C-ion (dotted). Kinetic curves: e, left - F_{ir} (T), F_{ir}^{max} (T_m~0.95) ; f, right - F_{iro} (T). IEx : *R*A/(B + C). Diffusivities - D_A: $\mathbf{D_B}$: D_C = 1: **5**: 0.2.

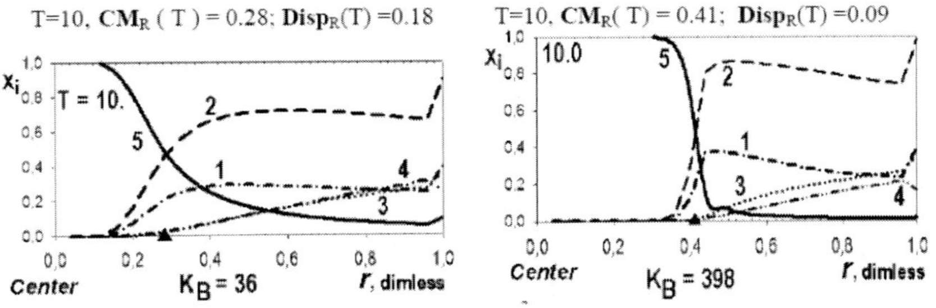

(*a*) (b)

Figure 3(a,b). (Variant 1. 5-components, Anion SO42-). Comparison of concentration waves - Xi(r,T) inside NC matrix of r-bead. Influence of KB value : KB=36(a, left) ; KB=398(b, right). Triangles (black) - CMR positions for 5R0 – waves (solids, 5) at r axis (abscissa). T=10; D3HSO4=0.0085<D1SO4=0.01<D4H=0.03.

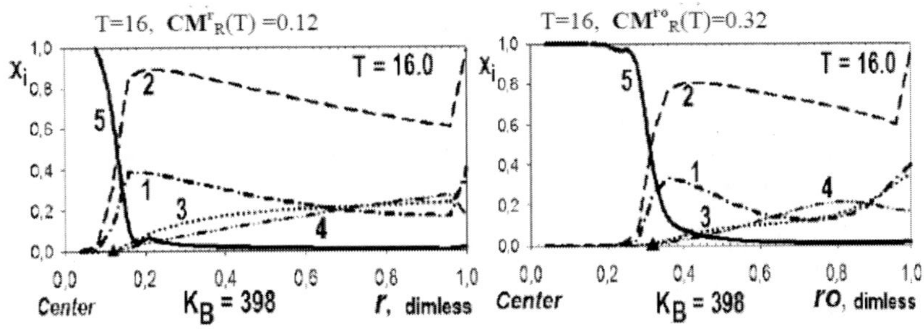

(*a*) (b)

Figure 4(*a*,b). (*Variant1. Anion* SO_4^{2-}) Comparison of $_5\boldsymbol{R}^0$ - waves (solids) inside the NC matrix of various shape: X_i(r,T), **r**-bead (left, *a*); X_i(ro,T), **ro**-fiber (right, b). Triangles (black) - $\mathbf{CM}_R$ positions at **r**, **ro** axes (abscissa). T=16; K_B= 398.

The waves propagation, as well as their concurrent interference–interaction should be used for the simulation and consideration of all the details of the mass transfer kinetics and dynamics. The fundamental monograph [13] (including additionally a lot of chromatographic papers by prof. F. Helfferich (USA) is devoted to the all aspects of the travelling "concentration waves" conception. There are a lot of research papers adjoined, which are included into the special issue [14] devoted to prof. F. Helfferich (Festschrift) with the detailed consideration of the travelling "multicomponent concentration waves" concept [14-19], including some others related.

A fairly detailed discussion of the wave concept, as well as many fundamental examples and definitions, are presented in a number of books and reviews (Refs. in [13-20]). In the presence of the effects of the non-ideality of the sorption medium the travelling concentration waves propagate with their broadening, including results of their consequent interference [13-25]. Naturally that the "multicomponent concentration waves" concept is widely used here for the consideration of the results of the computer simulation of the multicomponent kinetics in the bi-functional NC matrix. The examples of the multicomponent concentration waves are presented here in Figures 3-5 for the NC bi-functional matrix.

Comparison of the concentration waves behavior for the kinetic and dynamic system shows compatibility and differences for the interfering waves: there is the similarity between mass balance equations for multicomponent kinetics and dynamics. The travelling multicomponent concentration waves are originated in both cases.

However in addition, there is the difference:

a) In the *dynamic* IEx systems, as a rule, the travelling concentration waves used to reach the stage (named by F. Helfferich as "coherence" [13, 16]), in which the linear dimensions of the system are large, while multicomponent waves achieve the so called "coherent state" [13, 14, 16-18]. In this state the waves are separated by the concentration plateaus [13-20, 24, 25].
b) In the *kinetic* diffusion multicomponent IEx systems, the behavior of the waves and its interpretation differ in the presence of the diffusion phenomenon (even in the absence of reactions). Concentration waves in the kinetic systems propagate without the formation of the concentration plateaus between the waves [15]. In this case, due to a limited small size of the matrix the concentration waves are unable to disperse with the formation of concentration plateaus. Therefore, the coherent state is unattainable in these kinetic variants of mass transport. However, in the case of multicomponent kinetics, the effect of the interference of the multicomponent diffusion kinetic concentration waves still takes place [8, 15] Further, this behavior of the multicomponent system with wave interference are shown in Figures 3-5 based on the approach with the multicomponent kinetic equations of the mass transport in the bi-functional NC matrix.

The quantitative description of the waves behavior is successfully described here by the well known two integral parameters used for the characterization of any distributions: "Center of Mass" ($\mathbf{CM}_k$) and "Dispersion" ($\mathbf{Disp}_k$). The "$\mathbf{CM}_k$" parameter describes the integral "Center" of the frontal k-concentration distribution-wave, and the "**Disp**$_k$" parameter describes "the integral width" of the k-wave distribution. "**CM**" and "**Disp**" parameters are well known in the mathematical statistics as "average of distribution" and "dispersion" correspondingly [26]. Such types of the descriptions are used (Section 4, **CM**$_R$; **Disp**$_R$, Figures 3-6) for the estimation of the "completion Time"(T_{fin}) of the kinetic process and "width" of the $_k\boldsymbol{R}^0$-concentration frontal k-wave in Figures 3-6 ($_k\boldsymbol{R}^0$, solids, curves, 5). The concentration waves propagate across the various NC medium (**L**-membrane, **ro**-fiber, **r**-bead) from the boundary points (L_0, ro_0, r_0 = 1) to the final "zero points": L, ro, r = 0 correspondingly, Figures 3(*a*,b); 4(*a*,b); 5(b-d). The kinetic processes are finished at the completion moment T = T_{fin} , when the relation **CM**$_R$ + **Disp**$_R$ =1 is realized (Figures 6*a*,b). The more detailed description and discussion of the integral parameters will be given (Sections 4 and 5).

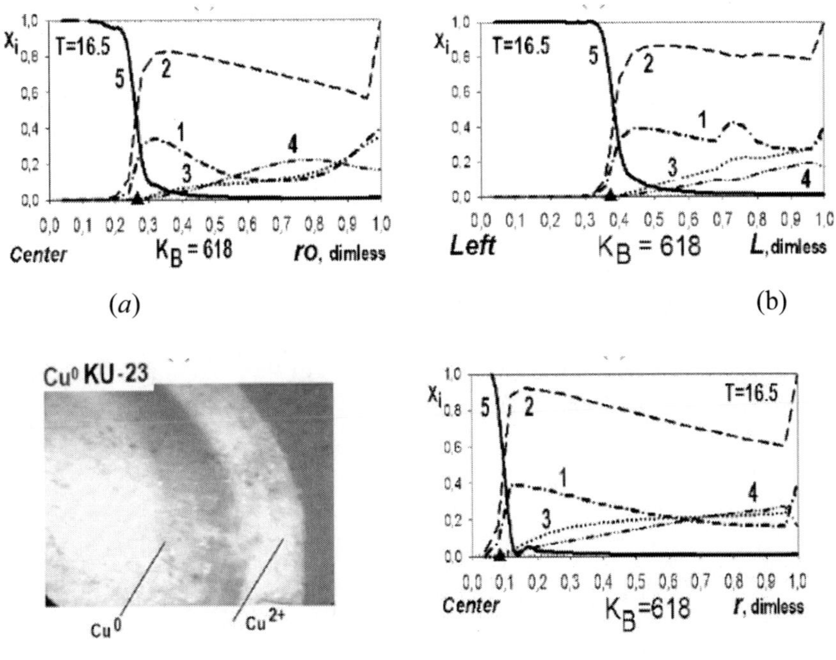

Figure 5 (*a*-d). Propagation of sharp concentration waves ("shell-core"type): (*a*) - microphotography of experimental example for oxidation process (visual Cu wave-boundary) inside KU-23 resin ; (b-d) theoretical calculated examples of sharp $_5R^0$ – concentration waves (5, solids): Xi(r, T); Xi(ro, T); Xi(L, T). Triangles - $CM^{r,ro,L}$ positions of the $_5R^0$– waves. K_B=618, T=16.5 (*Variant 1*).

The description of the created Model has been followed by the computerized simulation of the multi-component system behavior in the bi-functional NC matrix. All the parameters of the system in the computer simulations are taken as dimensionless (see Nomenclature) as it should be for any modeling. The obtained results of the computerized simulation for the multi-component mass transfer in the NC are original and new. The results of modeling are demonstrated and discussed in Sections 4 and 5.

The results of the simulation are presented here by a lot of various pictures (Figures 2-6). Additional illustrations may be shown visually via author's computerized animations prepared for the oral presentations (or by lap-top for posters). The animations are framed by the really calculated multicomponent concentration profiles for the successive time moments. The propagations of the travelling concentration waves in the NC matrix via animations may be demonstrated visually. Such type of the visual representations of the theoretical results has been using by author repeatedly including the Sessions of "IEX 2004, IEX 2008"Conferences [24, 25]. The animations prepared are perceived easily.

4. Concentration Waves in IEx and NC Kinetics

The described postulates of the mathematical Model were realized with taking into the consideration of all the relationships including multicomponent mass transfer kinetic differential equations, electro-neutrality relations, classical Nernst - Plank equations for the fluxes of the components and Mass Action Laws for the chemical reactions equilibrium (1) [2].

All the systems describing the multicomponent kinetic behavior in the bi-functional NC matrix of various shapes (**r**-bead, **ro**-fibre, **L**-membrane) were implemented for the simulation by author's computer programs. Several FORTRAN programs for such NC systems were arranged. The computer calculations were performed for a number of variants with different values of diffusion coefficients and various constants of the MAL chemical reactions (1), (Figure 1e). The results of the simulations for the multicomponent NC systems are presented. All the computer calculations have been obtained by using dimensionless values, including concentrations (X_i); diffusion coefficients D_i ; constants of chemical association-dissociation reactions (K_S). The mass transfer kinetics in the NC Model include the multicomponent concentration waves propagation along the dimensionless distance: **r**, **ro**, or **L** in the course of the dimensionless time (T). All the results of the computer simulation are new and obtained for the first time by author's modeling.

4.1. Computerized Simulation of Three (3)-Component Diffusion Kinetics in IEx Matrix

At first the three component diffusion kinetics with the concentrations [X_i], (i=A,B,C) for IEx system: RA/(B+C) inside the IEx matrix are considered. In this particular case there is no reactions and thus the IEx matrix is usual but no bi-functional. So in this case it is considered rather simple 3-ionic kinetics in the IEx matrix with three waves propagation in the course of time (Figures 2*a*-d). The boundary and initial conditions correspond to the IEx: RA/(B+C), where A,B,C - ions of three various kind: B, C-ions are invading into the matrix and A-ions in the resin are coming out of it (Figures 2*a*,b). The interference of the three A,B,C-concentration waves (Figures 2*a*-d) occurs in the course of 3-component diffusion for each matrix shape: in **r**-bead, **ro**-fiber or in planar **L**-membrane. Two Figures 2e,f show the corresponding kinetic curves in **r**-bead and **ro**-fiber: $F_{i\mathbf{r}}$(T),(e) and $F_{i\mathbf{ro}}$(T),(f).

The obtained results of the simulation show the interesting effect of the interference of the A,B,C- concentration waves, which should be marked: there is the *accumulation* effect (but only for the **r**-bead matrix) for the fastest B-ion wave moving and reflecting at the Center region (at r=0, Figure 2c).

The interference of the three i-concentration waves occurs in the course of three component diffusion (Figures 2*a*-d) in the **r**-bead or in the cylindrical **ro**-fiber. Figures 2 (**r**-bead, *a*,c, left ; **ro**-fiber, b,d, right) show the propagation of the i-component waves, which is dependent on the values of the diffusion coefficients D_i (i=A,B,C) and on the shape of the matrix. There is the considerable displacement effect with *accumulation* of the most mobile **B**-ions wave ($D_{\mathbf{B}}$>D_A>D_C) for the **r**-bead (Figure 2c). The significant displacement effect for the most movable B-wave (solid line, Figure 2c, T=1.1, left) takes place only for the **r**-bead. This accumulation is demonstrated visually via the author's computerized animations.

Thus under the certain conditions (i.e., for the big difference in diffusivities D_i, $D_{\mathbf{B}}$:D_A:D_C= **5:** 1: 0.2) the **B**-ions are accumulated very strongly at the Center of the **r**-bead (solid, Figure 2c). For this IEx process in the **r**-bead this accumulation effect brings the appearance of the kinetic maximum - $F_{i\mathbf{r}}^{max}$ (T_m ~ 0.95; Figure 2e, left) with the "non-standard" (non monotonic) $F_{i\mathbf{r}}$ -curve (Figure 2c).

For the **ro**-fiber the interference of the waves doesn't bring the **B** ions accumulation (Figure 2d). Therefore the kinetic $F_{\mathbf{iro}}(T)$ curve has the usual typical monotonic view (Figure 2f). The explanation for the B-ions accumulation (Figure 2c) with the $F_{\mathbf{ir}}^{max}$ for the kinetic curve (Figure 2f) is rather simple: due to the B-wave propagation the B-ions diffuse (along **r** - radius) into the fewer volumes of the sphere (**r**-bead) in the course of time T. This 3-component example shows the determining influence of the ion exchanger shape (**r** - sphere) of the porous IEx matrix.

Such particular case with various three diffusivities: D_A, $D_{\mathbf{B}}$, D_C was considered previously by other authors with getting only kinetic curves [6, 27, 28]. However the A,B,C - concentration waves-distributions inside the ion exchanger media (**r**-bead or **ro**-fiber, Figures 2 a-d) are original and have not been obtained previously.

The kinetic non monotonic curve $F_{\mathbf{ir}}$ (T) of the same type (with $F_{\mathbf{ir}}^{max}$) was obtained first during computer simulation of the 3- component IEx kinetics [6]. However propagating A,**B**,C-concentration waves have not been calculated previously. Besides explanations and reasons for the waves behavior with the obtained $F_{\mathbf{ir}}^{max}$ were not presented [6]. Kinetic curves with maximum for the multi-component IEx systems were obtained also on the basis of the another (macroscopic) model [27, 28]. However, that model [27] gives no possibility to calculate the A,**B**,C – concentration waves inside the IEx matrix. The reasons for the kinetic maximum were presented [27] qualitatively on the basis of that (macroscopic) model without the concentration distributions.

The results in this section show the application of the model to the more simple particular variant than for the cases to which the NC multicomponent Model is assigned. Therefore it should be noted here, that there is much more important for the created NC Model to get results in the computer simulation of the IEx kinetics inside the bi-functional NC matrix. The results of such calculations for the bi-functional NC matrix are presented in the next Section.

4.2. Simulation of Multicomponent Diffusion Kinetics with Additional Mass Transfer at the Active Nano-Sites ($_{k}R^{0}$) inside NC Bi-Functional Matrix

There is very productive for the quantitative estimation and description of the travelling concentration k-wave behavior to use the integral characteristics well known from mathematical statistics [26]: **C**enter of **M**ass - $\mathbf{CM}_k(T)$ and

Dispersion -$\mathbf{Disp}_k$(T). For the computer modeling these integral $\mathbf{CM}_k$(T), $\mathbf{Disp}_k$(T) parameters may be easily calculated by a computer during simulation of the process. The $\mathbf{CM}_k$(T) - dependence describes the velocity of the k- wave along the distance: radius for **r**-bead and **ro**-fiber or through the thickness (L) of the planar **L**-membrane. The $\mathbf{Disp}_k$(T) dependence describes the change of the width of the frontal X_i concentration wave in correspondence with its physical meaning. The illustrations are presented in Figures 3(*a*,b), 4(*a*,b) and 5(b-d) for the k-wave profiles with the $\mathbf{CM}_k$ positions. It is indicated by the triangles on abscissa **r**, **ro**, or **L** (Figures 3-5). The results of the computer calculations of the both integral parameters are presented in Figures 6(b-d).

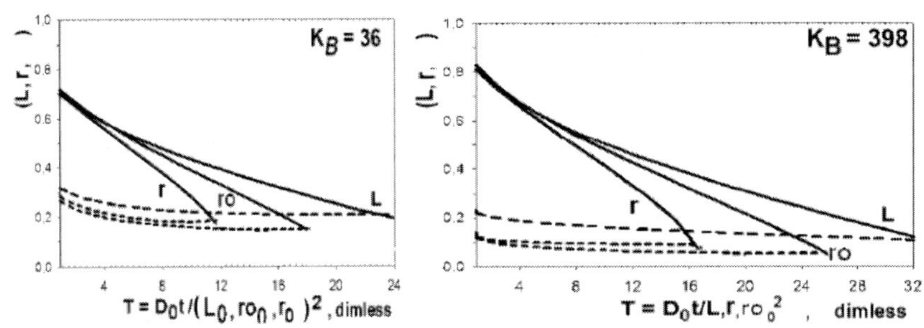

(*a*) (b)

Figure 6. (a,b). (*Variant 1.* Sulphate Anion: $_1SO_4^{2-}$). Time dependence of integral parameters: $\mathbf{CM}_R^{r,ro,L}(T)$, (solids)and$\mathbf{Disp}_R^{r,ro,L}(T)$,(dashed) during propagation of $_5\boldsymbol{R}^0$- wave : **L**-mem-brane (L), **ro**-fiber (ro), **r**-bead (r). D_{1SO4}=0.01 ; D_{3HSO4}=0.0085; D_{4H}=0.03. K_B = 36(*a*); 389(b). Ordinate axis: distance from boundary (L_0, r_0, ro_0 =1) to "zero"points (L, r, ro = 0):0< L, r, ro<1.

It will be shown in this investigation that the time dependences of this two parameters: $\mathbf{CM}_k$(T), $\mathbf{Disp}_k$(T) are very suitable for the description of the behavior of the $_k\boldsymbol{R}^0$-component wave, where $_k\boldsymbol{R}^0$ are nano-sites ($\boldsymbol{R}^0$ - example in Figure 1e). Such $_k\boldsymbol{R}^0$ - wave (curves 5, Figures 3-5, solids, 5) is formed in the bi-functional NC matrix due to the transformation of 4th-component ($_4H^+$ + $_k\boldsymbol{R}^0 \leftrightarrow _j\boldsymbol{R}H^+$) at $_k\boldsymbol{R}^0$ nano-sites with the change of [$_k\boldsymbol{R}^0$] - concentration in the k-wave (k = 5, *Variant 1,* Figures. 3-6, or k =6, *Variant 2*). Examples of such description will be presented in this Section 4.2 for *Variant 1* with two (1A,1B) chemical reactions and *Variant 2* with three chemical reactions (2.1)-(2.3) correspondingly. The lists of the components for each *Variant* in the NC matrix are shown below.

4.2.1. Sorption Kinetics of H_2SO_4 acid inside the bi-functional NC matrix with one reaction at the active nano-sites ($_5\boldsymbol{R}^0$); five (5) component system.

Variant 1. Kinetics of H_2SO_4 acid for the 5 - component NC System. The i-components (component number is the index to the left of the Symbol) with their corresponding diffusivities D_i :

$_1SO_4^{2-}$ – anions of acid (sulphates) with the diffusivity (D_1 =0.01) in the NC pores (2, Figure 1c) ;

$_2RH^+$ – immovable 2nd-component, formed by $_5\boldsymbol{R}^0$ nano-sites, protonated by acid cations ($_4H^+$) due to the association reaction (1B), (D_{2RH+}=0) ;

$_3HSO_4^-$ – anions of the acid (reaction (1A)) with the diffusivity (D_3=0.008) inside NC pores;

$_4H^+$ – cations of H_2SO_4 acid with the diffusivity (D_{4H+}=0.03) in the NC pores ;

$_5\boldsymbol{R}^0$ – zero valent nano-sites (D_{5R}=0), generated by the NP agglomerates (Figure 1e).

The IEx, NC system is characterized by n=5 components, three diffusion coefficients (D_{1SO4}; D_{3HSO4}; D_{4H+}) with two reactions: in the NC pores (1A) and at active nano-sites $_5\boldsymbol{R}^0$ (1B) correspondingly :

for NC pores, $_3HSO_4^- \leftrightarrow {}_4H^+ + {}_1SO_4^{2-}$;

$$[_3HSO_4^-] = K_A * [_4H^+] * [_1SO_4^{2-}] \text{ , } (MAL_A) \qquad (1A)$$

at active nano-sites $_5\boldsymbol{R}^0, {}_2RH^+ \leftrightarrow {}_5\boldsymbol{R}^0 + {}_4H^+$;

$$[_2RH^+] = K_B * [_5\boldsymbol{R}^0] * [_4H^+] \text{ , } (MAL_B) \qquad (1B)$$

The second relations in (1A), (1B) are presented by MAL_S with the constants: K_A, K_B . There is described in *Variant 1* the sorption of H_2SO_4 acid with sulphate anions (An^- =$_1SO_4^{2-}$) including concentrations [X_i] of all five components (i=1,2,…5; left index): [$_1SO_4^{2-}$], [$_2RH^+$], [$_3HSO_4^-$], [$_4H^+$], [$_5\boldsymbol{R}^0$]. Such diffusion five-component process in the bi-functional NC matrix is accompanied by two MAL_S relations: one reaction in pores (1A), with one more another reaction at the NC nano-sites (1B); (D_{1SO4}; D_{3HSO4}; D_{4H+}) - diffusivities of 1,3,4 - components.

4.2.2. Sorption kinetics of HCl acid inside the bi-functional NC matrix with two reactions at the active nano-sites ($_6\boldsymbol{R}^0$); six (6) component system

Variant 2. Kinetics of HCl acid for the 6 - component NC System. The i-components (the indexes are to the left of the Symbol) with their diffusion coefficients D_i :

$_1Cl^-$ – anions of acid (chlorides) with the diffusivity ($D_{1Cl} = 0.01$) in the NC pores (2, Figure 1c) ;

$_2RH^+$ – immovable 2nd-component, formed by $_6\boldsymbol{R}^0$ nano-sites protonated by acid cations ($_4H^+$) due to the association reaction (2.2), $D_{2RH+} = 0$;

$_3RHCl$ – immovable 3d-component, generated by the reaction (2.3), $D_{3RHCl} = 0$;

$_4H^+$ – cations of HCl acid, with the diffusivity (D_{4H+}=0.03) in the NC pores ;

$_5HCl$ – acid in dissociation-association reaction (2.1), the diffusivity D_5 =0.008 in the NC pores ;

$_6\boldsymbol{R}^0$ – zero valent nano-sites generated by the NP agglomerates (Figure 1e), D_{6R}=0.

The IEx, NC system (2.1-2.3) is characterized by n=6 components, three diffusion coefficients (D_{1Cl} ; D_{4H+} ; D_{5HCl}) with three reactions: one in the NC pores (2.1) with additional two reactions (2.2) , (2.3) at the active nano-sites $_6\boldsymbol{R}^0$ correspondingly:

for NC pores, $_5HCl \leftrightarrow {}_4H^+ + {}_1Cl^-$;

$$[_5HCl] = K_1 * [_4H^+] * [_1Cl^-] \rightarrow (K_1, MAL_1) , \qquad (2.1)$$

at active nano- sites $_6\boldsymbol{R}^0$, 1st step: $_2RH^+ \leftrightarrow {}_6\boldsymbol{R}^0 + {}_4H^+$;

$$[_2RH^+] = K_2 * [_6\boldsymbol{R}^0] * [_4H^+] \rightarrow (K_2, MAL_2) , \qquad (2.2)$$

at active nano-sites $_6\boldsymbol{R}^0$, 2nd step: $_3RHCl \leftrightarrow {}_2RH^+ + {}_1Cl^-$;

$$[_3RHCl] = K_3 * [_2RH^+] * [_1Cl^-] \rightarrow (K_3, MAL_3). \qquad (2.3)$$

The second relations in (2.1), (2.2) and (2.3) are described by MAL_S with the corresponding constants: K_1 , K_2 , K_3 There is described in *Variant 2* the sorption of HCl acid with chloride anions ($An^- = {}_1Cl^-$) including concentrations $[X_i]$ of all six components (i=1,2,…6; left index): $[_1Cl^-]$, $[_2RH^+]$, $[_3RHCl]$, $[_4H^+]$, $[_5HCl]$, $[_6\boldsymbol{R}^0]$. Such diffusion six-component process in the bi-functional NC matrix is accompanied by three MAL_S relations (2.1)-(2.3); D_{1Cl}; D_{4H+} ;D_{5HCl} are the diffusivities of 1,4,5 - components. The effect of two last "association-dissociation" MAL_S reactions (2.2), (2.3) may be easily transformed (by multiplication) and expressed as effective constant (product) K_{eff}

$$K_{eff} = K_2 * K_3 = [_3RHCl] /([_6\boldsymbol{R}^0] * [_4H^+] * [_5HCl]) \qquad (2.4)$$

Taking the relationships: (1A,1B) and (2.1-2.4) into account with comparing *Variant 1* and *Variant 2* it is clear that from two considered IEx, NC systems: *Variant 2* is more complex and has the ability for more variations (more degree of freedom) than *Variant 1*. All relations for the association-dissociation reactions (1A,1B) or (2.1-2.3) are presented here in the simple MAL_S type : $K_S = [ij]/[i][j]$ (Figure 1e).

5. Waves Behaviour in the Modelling of NC Kinetics

In this Section there are presented the results of the computerized simulation for the *Variants 1* and *2* with emphasis on the ${}_5\boldsymbol{R}^0$concentration waves propagation for the *Variant 1* (sorption of H_2SO_4 acid, 5 components, Figures 3-6). The description of the propagating ${}_6\boldsymbol{R}^0$ waves behavior for *Variant 2* is in the process of modeling. In Figures 6 the dependencies of the integral parameters: $\mathbf{CM}_R(T)$, and $\mathbf{Disp}_R(T)$ for the ${}_5\boldsymbol{R}^0$ – wave are presented for *Variant 1*, including various shapes of the NC matrix: **r**-bead, **ro**-fiber, **L**-membrane.

Comparison of the ${}_5\boldsymbol{R}^0$ - waves (5, solids) in Figures 3(*a*,b) shows that with the small K_B value (*a*, left) the ${}_5\boldsymbol{R}^0$- wave propagates to the Center faster, than with the large K_B value (b, right): $\mathbf{CM}_R$ *(left)* < $\mathbf{CM}_R$ *(right)*. For the dispersion the inequality is reverse:

$\mathbf{Disp}_R$*(left)* > $\mathbf{Disp}_R$*(right)*. In other words ${}_5\boldsymbol{R}^0$ - wave propagates slower with the profile, which is sharper (b, right) in comparison with the ${}_5\boldsymbol{R}^0$ - wave to the left (Figure 3*a*).

If K_B is increased (from (*a*) to (b)), it means that the equilibrium of the reaction (1B) shifts to the left hand side of the reaction (1B). In this case (from (*a*) to (b)) it is clear that such shift of the equilibrium (1B) brings the slowing-down of the kinetic process.

Figures 3(*a*,b) illustrate the influence of the K_B value in the reaction (1B): the increase of the K_B brings the increase of the completion time (T_{fin}) for the kinetic process.

Figures 4(*a*,b) illustrate the influence of the two various shapes of the NC matrix: **r**-bead (*a*, left) and **ro**-fiber (*b*, right) on the ${}_5\boldsymbol{R}^{0-}$ waves behavior for the same K_B value (K_B =398). Comparison of ${}_5\boldsymbol{R}^0$ – waves between 4*a*, (left) and 4b, (right) shows that for NC matrix of **r**-bead the ${}_5\boldsymbol{R}^0$ – wave (solid, curve 5) moves much faster than for the NC matrix of **ro**-fiber: $\mathbf{CM}^{\mathbf{r}}_R(T)$ (*a*) <

$\mathbf{CM}^{ro}{}_R(T)$ (b)) (compare the positions for CM_R (triangles) in Figures 4*a* and 4b).

The reason of such difference in the wave velocity is geometric : for **r**-bead the diffusion occurs in the fewer volume in the course of time (for **ro**-fiber the changes of the volume are not so steep).

The solid curves, 5 in Figures 5(b-d) represent the visual analogy to the experimental very sharp wave profile - Cu^{2+}/ Cu^0 (Figure 5*a*). The micrography (Figure 5*a*) was obtained by prof. T. Kravchenko group (Voronezh State University) [1] for oxidation process inside the **r**-bead of the domestic resin KU-23. Numbers (1-5) at the curves of Figutres 5(b-d) represent the numbers of the components (1,…,5) for *Variant 1* (Section 4.2.1).

The concentration wave propagation process in the **r**-bead may be expressed visually by the microphotography as it is shown in Figure 5*a*. The experiment is fulfilled in the prof. T. Kravchenko group (Voronezh University). Such type of the very sharp wave propagation inside the **r**-bead is well known in IEx kinetics and expressed by the term: “shell-core” kinetics [9, 12].

The comparison of the three Figures 5(b-d) for $_5R^0$ - wave shows that for large scale K_B value (K_B =618) in all three cases the very sharp waves (curve 5, solids) propagate along the radius (r, ro) or thickness (L) to the “zero points” (r,ro,L=0) of the matrix. The fastest sharp wave occurs inside the **r**-bead : (CM^r = 0.1), then in the **ro**-fiber (CM^{ro}= 0.27) and the slowest one inside the **L**-membrane: (CM^L= 0.37). So it is evident that for the completion time T_{fin} the inequality: $T^r{}_{fin}<T^{ro}{}_{fin}<T^L{}_{fin}$ takes place. The explanation is given below.

Figures 6 represent the time dependent estimations - L,ro,r (T) of the multicomponent NC kinetics for the three various shapes of the NC matrix. It is clear (see also Section 3) that the completion time (T_{fin}) of the kinetic process corresponds to the criterion with the relation: the amount of sum (**CM** + **Disp**)$r^{r,ro,L}$ = 1. In other words, distance covered by the $_5\boldsymbol{R}^0$ concentration wave to the moment T_{fin} is equal to the corresponding length of the matrix: L_0(thickness), ro_0(radius of cylinder), r_0(radius of sphere). In other words the completion time - T_{fin} corresponds to the “crossing” of the two curves: $\mathbf{CM}_R$(T),(solids) with $\mathbf{Disp}_R$(T),(dashed) (Figures 6*a*,b). In accordance with this criterion of “crossing”, there is obvious from all Figures 6(*a*,b) that for any conditions the kinetic mass transfer process in the NC matrix is the fastest inside **r**-bead and the longest inside **L**-membrane. The **ro**-fiber has the intermediate position in this row: $T_{fin}^{r} < T_{fin}^{ro} < T_{fin}^{L}$. The conclusions are obvious from Figures 5, 6.

6. Discussion

The computerized simulation of the mass transfer kinetics inside the NC matrix (Sections 4,5) discovers the clear analogies of the multicomponent NC kinetics with the main theoretical basis of the theory of nonlinear multicomponent chromatography [13, 18] These analogies concern the multicomponent concentration waves behavior in the bi-functional NC matrix. The physical meaning of the reasons for the explanation of the propagating concentration waves behavior (Figures 3-6) is described by the joint influence of the two coexisting factors (in the bi-functional NC matrix): Chemical Reactions (**I**) and Diffusion (**II**) (Figure 1e). Such type of the same joint influence in the theory of chromatography is represented by: the multicomponent isotherms for the equilibria (***I***) and set of the broadening factors for the concentration waves in columns (***II***), (i.e. well known effective HETP value for chromatographic columns) [8, 13, 18, 19].

The influence of the first (**I** - Reaction) factor is expressed by the K_S (MAL) values: the more is the K_B value (1B) – the less is the Dispersion (**Disp**$_R$) of the $_5\boldsymbol{R}^0$concentration wave. In other words, the width of the wave profile becomes narrow with the large K_B value : Figures 3(b)- 6(b-d). The comparison of **Disp**$_R$: (Figure 3*a,* left) with (Figure 3b, right), (solids, 5) shows it distinctly. The same effect is shown via the comparison of Figure 6a with Figure 6b (dashed, corresponding curves). The influence of the second factor (**II** - Diffusion) gives, as usual, the widening of the concentration waves with the increase of the diffusivity.

It should be emphasized here the non-trivial, specific effect for the NC system: there is no diffusivity for the $_5\boldsymbol{R}^0$-component wave (D_{5R}=0) , nevertheless the propagation of the $_5\boldsymbol{R}^0$ - concentration wave takes place (Figures 3-6). The physical reason of such propagation is not the diffusivity (D_{5R}=0) but the chemical reaction (1B) influence: the transformation for the components ($_2RH^+ \leftrightarrow {}_5\boldsymbol{R}^0 + {}_4H^+$) with the following mass transfer for the [$_5\boldsymbol{R}^0$] – concentration.

One more interesting result with the evidence of the above marked analogy can be seen from Figures 6(*a*,b). In this cases it takes place the typical behavior of the **Disp**$_R$ in the course of Time (T, abscissa): the **Disp**$_R$ of $_5\boldsymbol{R}^0$-waves tends asymptotically to the permanent value at the end of T- abscissa

(Figures 6(*a*, b), dashed curves behavior). In the theory of chromatography the same effect takes place for the concentration waves in column, when the favourable isotherm factor (***I***) used to compensate the unfavourable influence of the broadening factors (***II***) for concentration waves in columns: favourable equilibrium (***I***) compensate the widening of the concentration waves [13, 18, 19].

More detailed discussion will be implemented in further scientific journal publications. On the basis of the Model there are considered the results of the multicomponent kinetics inside the NC bi-functional matrix: for *Variant 1* (k=5) and *Variant 2* (k=6) (Section 4.2). The bases for the explanation and treatment of the $_k\boldsymbol{R}^0$ - concentration wave behavior are generated by the next rules.

Variant 1. Association – dissociation reaction (1B) at the nano-sites $_k\boldsymbol{R}^0$ brings the retardation (in the NC medium) of the i-component due to the partial transformation: $_k\boldsymbol{R}^0 + {}_4H^+ \leftrightarrow {}_2RH^+$ (1B). Therefore the movement and broadening of the $_k\boldsymbol{R}^0$ - concentration waves depend on the K_B value.

The comparison between Figure 3(*a*, left) and Figure 3(b, right) shows that the width of the $_5\boldsymbol{R}^0$-wave (5, solid, *a*, left) is much larger than for the $_5\boldsymbol{R}^0$-wave (5, solid, b, right). In addition the $_5\boldsymbol{R}^0$-wave (3*a*, left, solid, K_B=36) moves faster than the $_5\boldsymbol{R}^0$-wave (b, right, solid, K_B=398): $CM^a{}_R < CM^b{}_R$.

Variant 2. Association – dissociation two steps reactions (2.2), (2.3) at the nano-sites $_6\boldsymbol{R}^0$ bring the retardation (in the NC matrix) of the i-component due to the partial transformations: 1st step $_6\boldsymbol{R}^0 \leftrightarrow {}_2RH^+$ (2.2) and then due to the 2nd step: $_1Cl^- + {}_2RH^+ \leftrightarrow {}_3RHCl$ (2.3). The total influence of the two step reactions is determined by the value of the effective coefficient K_{eff} (2.4).

The corresponding computer animations are prepared by author for the visual demonstration of the obtained results of the modeling. The animations obtained on the basis of the computer calculations are perceived easily.

CONCLUSION

The modern kinetic Model for the multicomponent mass transfer in the novel NC materials is created. The computerized simulations of the nonlinear IEx, NC systems on the basis of the Model bring the new results describing the behavior of the multicomponent concentration waves in the bi-functional NC matrixes for the three various shapes: **r**-bead, **ro**-fiber, **L**-membrane.

The results demonstrate rather comprehensive analogy between theory of nonlinear chromatography and multicomponent kinetics in the bi-functional

NC matrix. This analogy is used for the description of the multicomponent concentration waves behavior. In the interpretation of the analogy there is shown the behavior of the multicomponent concentration waves in the bi-functional NC matrixes for the three various NC matrix shapes. In the interpretation of the analogy there are shown the decisive influences of the equilibrium parameters and diffusion effects on the propagation of the concentration waves during the mass transfer inside the bi-functional NC matrix.

The "multicomponent concentration wave" concept in the study of the bi-functional NC matrix with the two factors: (**I**) Reactions on the active nano-sites and (**II**) Diffusion in the NC medium is effective and brings the clear, and understandable treatment of the multicomponent mass transfer kinetics in the NC matrix. The quantitative estimations of the concentration waves behavior may be obtained by using two integral parameters of the wave distributions: "Center of Mass - **CM**", and "Dispersion - **Disp**". The corresponding conclusions concerning co-influence of the reactions and diffusion coefficients inside the bi-functional NC matrix on the ${}_{k}\boldsymbol{R}^{0}$ - concentration wave behavior are obtained.

The obtained original, new results show the behavior of the broadening and interacting propagating concentration waves inside the bi-functional NC matrix during multicomponent mass transfer kinetics. The corresponding computerized author animations are prepared for the visual demonstration of the computerized results of the simulation. The kinetic process regularities illustrated by the animations (framed on the basis of the computer calculations) are perceived easily.

Acknowledgments

Part of the research has been fulfilled in Karlsruhe, (KIT) under the support of DFG (Deutsche Forschung Gemeindschaft, 2011, German-Russian grant N 436 RUS 113/989/9 - 1).

References

[1] Kravchenko T., Polyansky L., Kalinitchev A. and Konev D., *Nano Composites Metal - Ion Exchangers*. M. Nauka (Sci.), 390P (2009). (Russian).

[2] Kalinitchev A., Kinetic and Dynamic Chromatographic Systems, and Models of Mass Transport: Behavior of Multicomponent Concentration Waves. *Protection of Metals and Physical Chemistry of Surfaces.* 47: 698-706 (2011). (in Russian and English: P.570-579).

[3] Kiprianova E., Kravchenko T., Konev D., Kalinitchev A. and Hoell W., Reducing sorption of the molecular oxygen from water by silver-sulpho-cation exchanger NanoComposite with various ionic forms., *Rus. J. Phys. Chem.*. 84: (N6) 1104 - 1110 (2010).

[4] Helfferich F. and Hwang Y-I., Kinetics of acid uptake by weak base anion exchangers. Mechanism of proton transfer. *Am.I.Ch.E. Symp. Ser.* 81,17-27(1985).

[5] Kalinitchev A., Kolotinskaya E. and Semenovskaya T. Computerized analyses of the diffusion processes in complexing ionites. *J. Chromatogr.*, 243, 17 - 24 (1982).

[6] Hwang Y-I. and Helfferich F., Generalized model for multispecies IEx kinetics including fast chemical reactions. *Reactive Polymers*, 5: 237-252 (1987).

[7] Petruzelli D., Helfferich F., Liberti L., Millar J., and Passino R, Kinetics of IEx with intraparticle rate control: Models Accounting for Interaction in the solid phase. *Reactive Polymers*, 7, 1-13 (1987).

[8] Kalinitchev A., Investigation of Intraparticle IEx Kinetics in Selective Systems, in (eds.: Ja. Marinsky and Y.Marcus) *IEx and Solv. Extr.* 12, M.Dekker, Ch.4: 149-196 (1995).

[9] Helfferich F., Ion Exchange Kinetics. Ch.5. in *Ion Exchange* (A series of Adv.) ed. Ja.I. Marinsky). St. Univ. NY.1 (1966).

[10] Haase R., Thermodynamics of Irreversible Processes, M.: "Mir" Ed. (1967). P.544 (Rus. Transl.)

[11] Helfferich F., *Ion Exchange*. Mc.Graw- Hill. NY. Ch. 6. 255- (1962). (*Ionen Austauscher*, Verlag Chemie, GMBH, (1959), Deutsch).

[12] Helfferich F., Ion Exchange Kinetics – Evolution of a Theory. in:"*Mass Transfer and Kinetics of IEx*" (ed: L.LibertiandF.Helfferich) M.Sijthoff and Nordhoff . The Hague. 157-179 (1983).

[13] Helfferich F. and Klein G., *Multicomponent Chromatography. Theory of Interference*, New York: M. Dekker Inc, (1970) 360P.

[14] *Industrial and Engineering Chemistry Research. Journ.* (prof. *F. Helfferich* Festschrift), Am. Chem. Soc. J. 34: 2551-2922 (1995).

[15] Tondeur D. and Bailly M., Unifying Concept in non-linear unsteady processes. PII. Multicomponent waves, competition and diffusion. *Chem. Eng. Process.* 22: 91-105 (1987).

[16] Tondeur D., Paradigms and Paradoxes in Modeling Adsorption and Chromatographic Separations. *Ind. and Eng. Chem. Res.*, 34: 2782-2788 (1995).

[17] Hwang Y.L., Wave propagation in Mass transfer Processes: from chromatography to Distillation. *Ind. and Eng. Chem. Res.*, 34: 2849-2864 (1995)

[18] *Percolation Process: Theory and Applications,* ed.: Rodrigues, A.E., Tondeur, D., The Hague: Sijthoff and Nordhoff, 1981.

[19] Kalinitchev A., Nonlinear Theory of multicomponent sorption dynamics and Chromatography. *Rus. Chem. Reviews* 65: 95-115 (1996) (in Russian and English).

[20] Hoell W. and Kalinichev A., The theory of formation of surface complexes and its application to the description of multicomponent dynamic sorption systems. *Rus. Chem. Reviews*, 73: 351-370 (2004) (Russian and English).

[21] Courant R. and Friedrichs K., *Supersonic flow and Shock Waves*, Springer (1999) P.464.

[22] Prigogine I. and Herman R. *Kinetic Theory of Vehicular Traffic* .NY.: Elsevier (1971)55P.

[23] Whitham G., Linear and Nonlinear waves., Wiley.NY. (1974).

[24] Kalinitchev A. and Hoell W., Multi-component Ion Exchange dynamics with the equilibria described by surface complexation theory, in *IEX2004, Ion Exchange Technology for Today and Tomorrow,* ed.: M. Cox, Soc. Chem. Ind., London, 53-58 (2004).

[25] Kalinitchev A. and Hoell W., Theoretical Principles of multicomponent frontal and displacement elution Chromatographyon on the basis of surface complexation theory, in *IEX 2008, Recent Advances in IEx TheoryandPractice,* ed: M. Cox, Soc. Chem. Ind., London, 85-93 (2008).

[26] Korn G. and Korn T., Mathematical Handbook, Ch. 18, 19 McGraw-Hill, NY-Sidney 1968.

[27] Dolgonosov A., Khamisov R., Krachak A. and Prudkovskiy A., Macroscopic model for multispecies IEx kinetics. *React. Func. Polymers.* 28:13-20 (1995).

[28] Dolgonosov A., Electric Effects in multicomponent IEx kinetics. *React. Func. Polymers.* 34: 47-51 (1999).

In: Smart Nanoobjects
Editor: Kirill Levine

ISBN: 978-1-62808-741-3

Chapter 7

DEPOSITION OF THIN FILMS OF NANOMATERIALS BY THE SURFACE ACOUSTIC WAVE ATOMIZER

O. V. Balachova, S. M. Balashov and A. Pavani Filho
Center for Information Technology Renato Archer – CTI, Campinas, São Paulo, Brazil

ABSTRACT

The surface acoustic wave (SAW) atomizer was used for deposition of thin films of nanomaterials, namely, carbon nanotubes and graphene oxide onto a flat silicon substrate. Atomization was made in "digital regime" (i.e. droplet by droplet). It was shown that the use of SAW atomization allows one to obtain submicron-thick films on a small surface. The thickness of obtained films was controlled by controlling a number of deposited droplets. It was observed that the SAW atomization can cause an additional exfoliation of bulk graphite oxide fragments.

INTRODUCTION

One of the rapidly developing fields of modern microelectronics is the development of chemical sensors. In traditional sensors, sensitive films have good adhesion to the surface of the sensor due to chemical bonding. These films can be obtained by a number of techniques which provide their good uniformity: sputter deposition, atomic layer deposition, CVD etc. Recent progress in synthesis of nanomaterials like decorated carbon nanotubes (CNT), graphene or graphene oxide (GO), among others, made it possible to obtain the amount of them significant enough to use them as materials for sensitive thin films deposition, which caused the appearance of the new class of sensors [1-3]. The big advantage of the films made of such a materials is that the stable response can be obtained even from the very small-area films. In contrast to traditional materials, the new materials have the adhesion strength determined mainly by short-range relatively weak forces, which makes traditional methods of film deposition ineffective.

Due to the high degree of saturation of chemical bonds in nanomaterials and the big size (in molecular scale) of their molecules, the most common form of the 'raw material" for the film deposition is the suspension of them in some liquid, making, thus, spray and spin coating the most promising techniques for deposition. To be used as sensitive elements, nanostructural materials should be deposited in the way to ensure the small area of coverage, small thickness and high uniformity of the film. In some cases the resulting film should occupy only a fraction of the surface of the circuit. Traditional methods of spray or spin coating have certain internal limitations, which strongly restrict their use for such cases. Traditional spray, for example, has the droplet size, which is too big, and even a minimum quantity of liquid, which could be transferred to the surface of the sample in a controllable way, provokes convective flows inside the liquid layer during the process of drying. The convection redistributes deposited material along the surface of the sample causing structural nonuniformities in the final film. The same is true for spin coating as well. The additional roughness of resulting thin films deposited by traditional methods comes from the fact that CNTs or graphene are used in the form of suspensions. Solid material distributed in suspensions tends to form agglomerates in convective flows, with nonuniform distribution of the size. This nonuniformity can modulate mechanical and electrical properties of the deposited film causing degradation of the sensor response.

In the present work, in order to overcome the difficulties mentioned above, we used "digital" spray, which allows one to nebulize and deposit onto various surfaces microlitric quantities of suspensions. The nebulizer was made as the symmetric SAW atomizer. The results of depositions of thin films of CNT and GO are presented. It was shown that thin films obtained by this method can be used as sensitive layers for surface acoustic wave sensors.

EXPERIMENTAL SETUP AND ATOMIZATION PARAMETERS

The SAW atomizer and its equivalent circuit are shown in figure 1. The droplet of suspension (4) for atomization was placed in the center between two identical Interdigital Transducers (IDTs) (2) printed on the surface of 127° YX-cut of the $LiNbO_3$ (5).

The continuous RF signal from the generator (1) was applied through the matching circuit (6) to IDTs. SAWs propagating from each IDT (3) to the center of the plate, reaching the droplet, atomized it, producing the spray of nanolitric droplets (7), which was deposited directly onto the surface of the sensor (8). Details of construction of the atomizer can be found elsewhere [4].

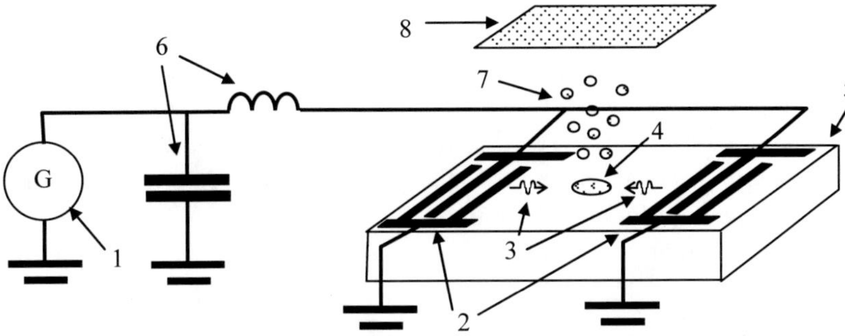

Figure 1. Schematic representation of the film deposition by the SAW atomizer. 1 – generator; 2 – IDTs; 3 – SAWs propagating from the IDT to the droplet; 4 – droplet of the nanomaterial suspension placed in the center of the crystal plate; 5 – piezoelectric crystal plate; 6 – elements of the matching circuit; 7 – spray of nanolitric droplets of the suspension; 8 – sample onto which the spray should be deposited.

This atomization took place in "digital regime", which means that submicrolitric droplets of suspension (4) were placed in the centre between IDTs and atomized one by one by continuous application of the RF signal to them. No additional fixture was used to keep the droplet still during atomization.

Typical droplet size distribution, obtained with the Spraytec system (Malvern Instrument Ltd., USA) is shown in the figure 2. The size distribution of droplets in the spray has maximum at 50 μm diameter, which, providing that only few droplets were atomized, guarantees submicron thickness of the resulting film.

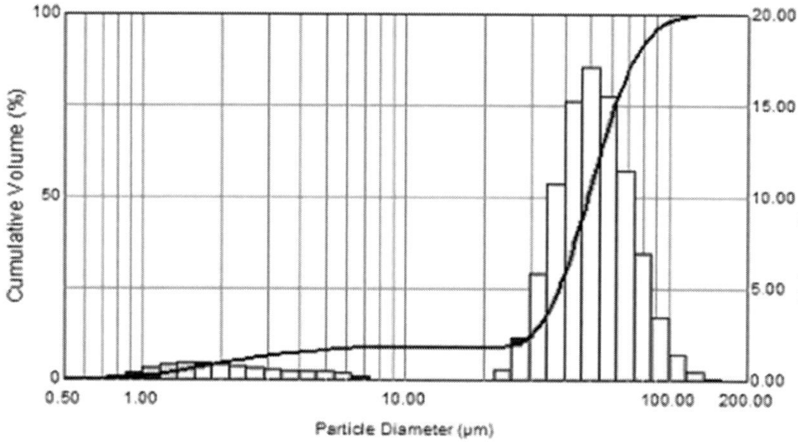

Figure 2. Typical particle size distribution in the spray of the atomized droplet.

The GO water suspension of concentration 0.33 mg/ml was used for atomization. Material for the GO suspension was obtained by oxidation of the bulk graphite with posterior sonification of graphite oxide in order to exfoliate it into two-dimensional flakes of the GO. After centrifugation the resulting suspension was used for atomization.

It was found that the final suspension of GO always had the fraction of the bulk graphite oxide. The volume of each atomized suspension droplet was 0.35 µl. The 0.137 mg/ml suspension of multiwall CNTs in NMP (N-Methylpyrrolidone) was used for atomization. In this case the size of suspension droplets was from 0.75 µl to 1.00 µl. In both cases the Si samples were used as the substrate for deposition of atomized nanodroplets.

Results and Discussion

During the process of atomization very strong gradients of mechanical forces caused by interference of the SAW with liquid appear inside the droplet intensively mixing its content. One of the consequences of this process is additional homogenization of the suspension. In addition it was found that the SAW appeared to be capable to split the bulk samples of the graphite oxide in suspension into two-dimensional GO flakes (see figure 3).

This result shows that the SAW can be used in future for additional exfoliation of very small quantities of nanomaterials. The structure of the deposited GO thin film is shown in the figure 4. The presence of well defined separation between deposited nanodroplets can be attributed to the fact that during atomization their additional charging takes place [5]

Figure 3. Exfoliated by the SAW samples of the graphite oxide deposited onto the Si surface by the SAW atomizer.

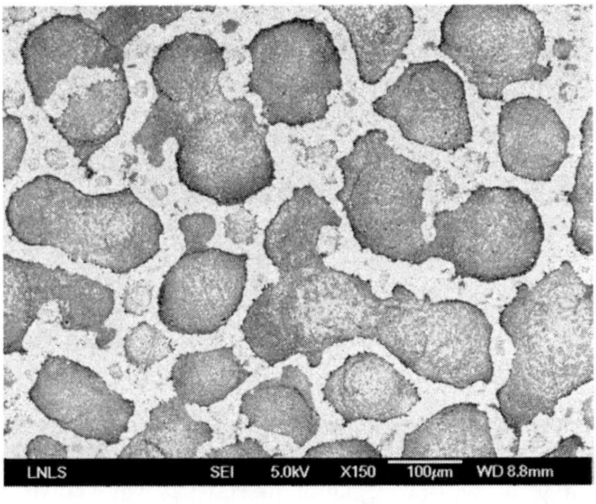

Figure 4. Thin film of the GO deposited on the Si surface by the SAW atomizer after atomization of five droplets of the suspension.

Due to a small number of atomized droplets the resulting surface coverage is not complete. This type of coverage was chosen to use the obtained GO film as the sensitive element for the SAW sensor, for which the most important parameter is the film thickness. Measured thickness of the GO film was in the range 10 - 100 nm, which allows one to use it for the SAW sensors with frequency up to 1 GHz. Note that the GO film in the figure 4 has uniformity much higher then in the case of free drying of the microlitric droplets. It happens because the process of drying up of nanolitric droplets of deposited spray does not accompanied with convective flows in liquid.

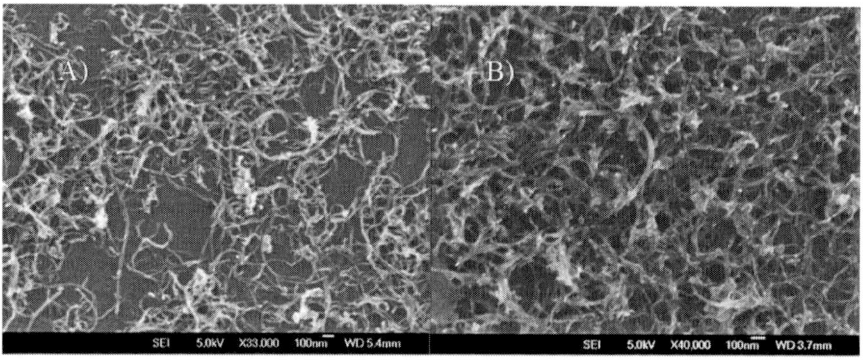

Figure 5. Results of deposition of the suspension of CNTs onto the surface of silicon. A – atomization of two droplets 1 μl each. B – atomization of five droplets 1μl each.

The result of deposition of CNT suspension onto the surface of silicon is shown in the figure 5. As it can be seen the proposed method of deposition allows one, atomizing the suspension droplets one by one, to control final thickness of the film even for very low thickness.

CONCLUSION

The process of atomization of microlitric droplets of suspension of nanomaterials was used to obtain low thickness nanomaterial films. It was shown that sonification of the suspension by the SAW causes additional exfoliation of bulk nanomaterial and can be used for future improvement of the quality of small quantities of nanomaterials. It was shown that droplet by droplet atomization can be used for control of the thickness of very thin films. Considering that the obtained films have thickness in nanometer range, it is possible to use them, for example, as sensitive elements of SAW sensors of gigahertz range.

ACKNOWLEDGMENT

The authors acknowledge the valuable financial support of the CI-Brazil project (CNPq, process no. 550495/2011-8), INCT NAMITEC (CNpQ, process no. 573738/2008-4/FAPESP, process no. 2008/57862-6) and PCI/MCT-Brazil. We also would like to thank Dr. Stanislav Moshkalev (CCS Unicamp, Brasil) for provided samples of the graphene oxide. We are grateful to the Brazilian Synchrotron Light Laboratory (LNLS)/C2NANO for the possibility to make the electron microscope (SEM-FEG) images of the obtained films.

REFERENCES

[1] V. Chivukula, C. Kritzinger, F. Yavari, D. Ciplys, N. Koratkar and M. S. Shur, in *Proc. 2010 IEEE Ultrasonics Symposium*, 257 (2010).

[2] Erdem, M. Muti, P. Papakonstantinou, E, Canavar, H. Karadeniz, G. Congur and S. Sharma, *Analyst*, 137 (9), 2129 (2012).

[3] J. D. Fowler, M. J. Allen, V. C. Tung, Y. Yang, R. B. Kaner and B. H. Weiller, *ACS Nano*, 3 (2), 301 (2009).

[4] O. V. Balachova, S. M. Balashov, A. Pavani Filho and M. G. de Almeida, *ECS Transactions*, 31 (1), 267 (2010).

[5] J. Ho, M. K. Tan, D. B. Go, L.Y. Yeo, J. R. Friend, H-C. Chang. *Anal. Chem.* 83, 3260 (2011).

In: Smart Nanoobjects
Editor: Kirill Levine

ISBN: 978-1-62808-741-3

Chapter 8

LIQUID PHASE EXFOLIATED MULTI-LAYER GRAPHENE: FABRICATION AND CHARACTERIZATION

A. V. Alaferdov, M. A. Canesqui and S. A. Moshkalev
Center for Semiconductor Components,
State University of Campinas,
Campinas, Sao Paulo, Brazil

ABSTRACT

Liquid phase exfoliation of natural graphite to produce high quality micron scale multi-layer graphene sheets (flakes) in two different solvents (n,n - dimethylformamide and isopropyl alcohol) has been studied. The effects of process parameters (basically, sonication time, bath temperature, and centrifugation time) have been investigated. By adjusting the process parameters, high-density solutions containing graphene sheets with lateral sizes from 0.5 to 10 μm (with lateral size-to-thickness ratios up to 200-300) were obtained. For characterization, graphene sheets were deposited over holey carbon TEM grids. The statistic analysis for large number of sheets was performed to show a log-normal lateral graphene size distribution. High quality of graphene sheets was confirmed using scanning electron microcopy and Raman spectroscopy.

INTRODUCTION

The separation of single-layer (SLG) graphene sheets in 2004 has called much attention of the research community due to extraordinary electrical, thermal, optical and magnetic properties of graphene [1-4]. Besides the SLG, at the moment multi-layer graphene (MLG) is also a subject of much interest [5]. MLG consists of a number of overlapping individual graphene monolayers that are bound by a week van der Waals force.

Electrons in SLG and MLG exhibit different behaviors. The electronic structure of MLG differs strongly from that of SLG due to an interlayer coupling [6]. Electrical, thermal and other properties have shown to change gradually with increasing number of layers *N*, the strongest changes occur when *N* changes from 1 to just 2-3. MLG has many potential applications in sensors, transistors, supercapacitors, fuel cells, solar cells, nanocomposites, etc. [7-12]. Fabrication of MLG with large lateral dimensions and high quality is currently a subject of intense research.

There are various methods for preparing MLG, such as chemical vapor deposition [13], super-short-pulse laser produced plasma deposition [14], reduction of multi-layer graphene oxide [15], exfoliation through thermal shock [16], acid treatment of expandable graphite [17], etc. All of these methods have certain merits and disadvantages.

Some of them are toxic, others are energy-intensive and time consuming, can produce a lot of defects in MLG, or are not usable for large-scale production. The method used in this work, exfoliation of graphite in solvents [18-25], is low-cost and allows obtaining a solution dispersed high quality graphene flakes.

EXPERIMENTAL SECTION

Materials

Natural graphite powder (size of polycrystals 1-3 mm) was purchased from Nacional de Grafite Ltda (Brazil). Two types of solvents were used: n, n-dimethylformamide (DMF) and isopropyl alcohol (IPA), both analytical grade.

Preparation

Graphene dispersion was prepared from mixture of natural graphite flakes (1 mg) in DMF (1 ml) or IPA (1 ml) using ultrasound processing followed by centrifugation. It is known that for effective process of exfoliation, the surface energy (Hildebrand solubility parameter) of solvents and solute must very close to each other [26-28]. In our case, Hildebrand parameters for DMF and IPA are 24.86 and 23.58 $(MPa)^{1/2}$ respectively (graphene ~ 23 $(MPa)^{1/2}$) [29]. Bath sonication was performed using a sonication bath Unique USC-1880 with power of 100 W and frequency of 37 kHz, with the time of sonication varying from 2min up to 8h. After sonication, the obtained dark dispersions were centrifuged using a MiniSpin Eppendorf AG 22331 with time and velocity varying from10 to 90 min and from 800rpm to 10 000rpm, respectively, in order to remove larger flakes of graphite and to separate graphene flakes of different sizes. After centrifugation, the top 0.4ml solution volume was carefully removed by pipette and retained for analysis and further use.

Characterization

In order to characterize and analyze the dispersion of natural graphite into graphene sheets (flakes), we deposited the obtained solutions contacting graphene onto (i) silicon substrate and (ii) standard holey carbon mesh grids (mean size of holes~1 μm) by drop casting, one to a few drops were used for sample deposition. The second method appeared to be more efficient for analysis, and was mostly used in this study. First we examined the samples by Scanning electron microscopy (SEM) with high resolution mode, and then the quality of graphene flakes was analyzed by micro-Raman spectroscopy in confocal configuration using an NT-MDT NTEGRA Spectra with 473 nm and 633nm lasers.

Exfoliation of Graphene

The exfoliation of graphite by sonication in a liquid phase is believed to be the result of shock waves and micro-jets generated in the sonicated liquid [30, 31]. Cavitation is a nonlinear phenomenon that concentrates and transforms a low-density, elastic wave energy into higher densities through rapid formation and collapse of gas bubbles in the liquid. The resulting extreme local conditions induce profound changes at the solid-liquid interface.

Collapse of the bubble on the surface itself will cause breaking by the shock waves produced, whereas collapse in the liquid close to as surface causes a microjet of solvent that can hit the solid with a great force [32, 33]. The mechanical energy released is sufficient to remove top layers after disrupting weak molecular interactions, leading to delamination and dispersion of the initial graphite flakes. It is suggested that a polycrystalline natural graphite starts to exfoliate in the areas containing defects and next at the grains boundaries. The schematic of the exfoliation process is shown in Figure 1. Suspensions of graphene sheets have been fabricated in DMF and IPA with different time of sonication (see in Figure 2 some examples of solutions).

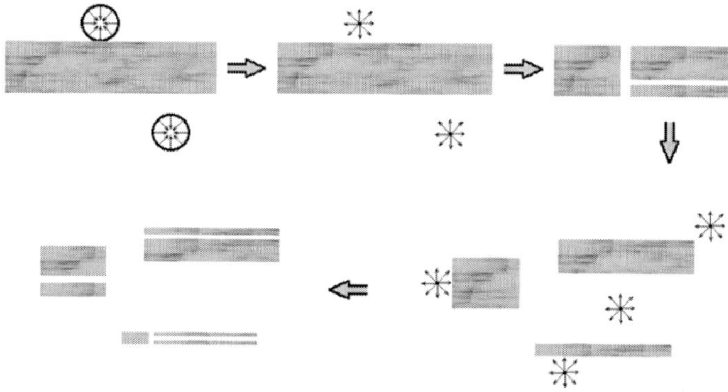

Figure 1. Schematic of the process exfoliation natural graphite in liquid phase by a cavitation process. Formation of bubbles and shock waves, followed by graphite delamination is shown.

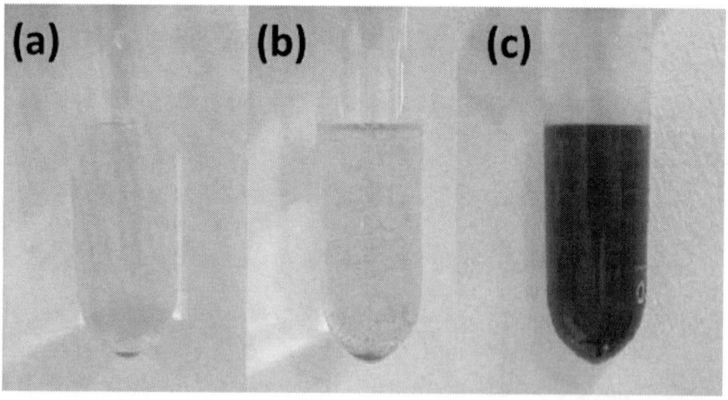

Figure 2. Solutions of MLG in DMF. (a) – mixture of DMF and natural graphite, (b) – suspension after 2 minutes of sonication, (c) – suspension after 240 minutes of sonication. Logarithm.

RESULTS AND DISCUSSION

Kolmogorov in 1941 reported that statistics of metal (gold) particles after mechanical processing (attrition) of the gold-contating rocks followed the so called log-normal size distribution, i.e., the statistics for the particle size logarithms is normal (Gaussian):

$$f(x) = \frac{1}{x\sigma\sqrt{2\pi}} exp\left\{-\frac{1}{2}\left(\frac{\ln x - \mu}{\sigma}\right)^2\right\} \quad (1)$$

where $L_C = exp(\mu)$ – median size,

$\langle L \rangle = exp\left(\mu + \frac{\sigma^2}{2}\right)$ – expected value,

$w = exp\left(\mu + \frac{\sigma^2}{2}\right)\sqrt{\exp(\sigma^2) - 1}$ - standard deviation.

This kind of distribution was observed later in many different situations, and can be expected to be valid in the our case. To analyze the size distribution, histograms of lateral sizes for graphene sheets deposited over holey TEM grids were first obtained, with the number of measurements always being high enough (>100) (Figure 3). To check the hypothesis of the log-normal distribution in the present case, we used the Pearson's chi-squared test [35, 36].

$$\chi^2 = \sum_{i=1}^{n} \frac{(O_i - E_i)^2}{E_i} \quad (2)$$

O_i – an observed frequency (or weight),

E_i – an expected (theoretical) frequency (or weight), asserted by the hypothesis,

n – number of intervals in the histogramm.

This test allows to estimate the probability that deviation of the experimentally measured distribution from the expected theoretical curve is caused by occasional reasons, and not by the error in the hypothesis. If the probability is high, then the deviation is treated as occasional thus confirming the hypothesis.

More details on this analysis can be found elsewhere [37]. The obtained results and analysis confirmed that the size distribution of graphene sheets produced in IPA and DMF solutions follows the log-normal law, as can be seen in Figure 3 (results for IPA not shown).

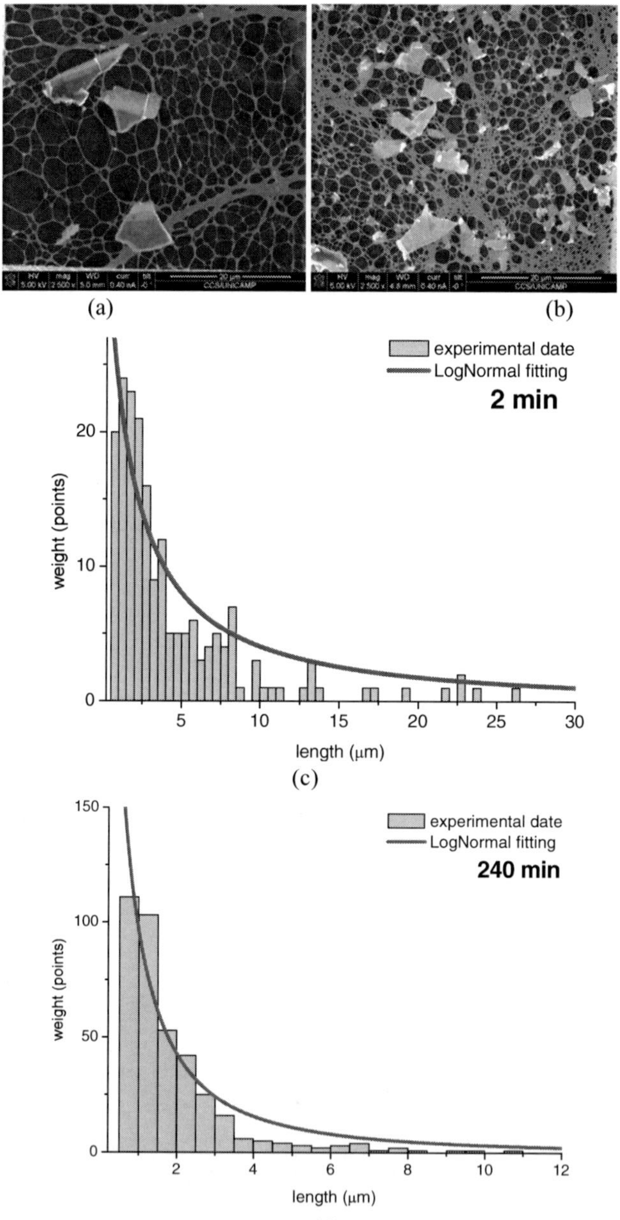

Figure 3. a-,b- SEM images of graphene sheets deposited from DMF solution on holey carbon grids, for different sonication times (2 and 240 min respectively); c-, d- statistical distribution of graphene lateral sizes and lognormal fitting to the histogram for 2 and 240 min. of sonication in DMF.

For short period of sonication, (2 min.) the probability that the distribution is described correctly by low-normal law, is estimated to be P>80%, while for longer sonication times this probability increases, being P>95% for 240 min.

The effects of different factors (time of sonication and centrifugation, type and temperature of solvent) on the size distribution of the graphene sheets were analyzed (Figure 3(a, b), 4). In the study of the effect of sonication time, it was found that during first 2 minutes graphene sheets with lateral size up to 10 μm are mostly formed (more than 50% of the sheets measured on TEM grids are those with size exceeding 10 μm). For longer time, the mean size decreases (to ~1 μm for 240 min. time).

Aspect ratio (ratio of lateral size to thickness) was estimated roughly using SEM images found to vary within a range 50-300 in all cases studied here, depending slightly on the sonication and centrifugation time and the type of solvent.

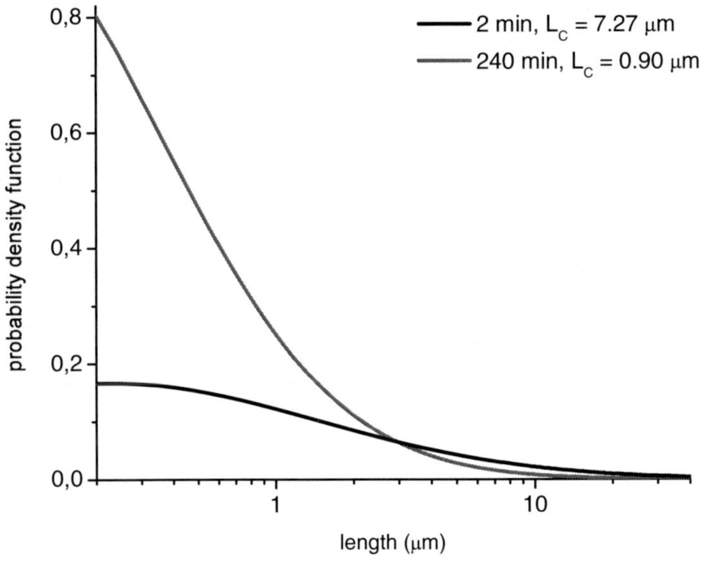

Figure 4. Distribution function (after fitting) of graphene sizes for 2 and 240 min. of sonication in DMF.

For study of the effect of centrifugation time, suspensions obtained with 2′ and 240′ sonication were used , see Figure 5. Centrifugation time was 15′ and 90′, and speed 800 rpm. For 2′ sonication (Figure 5c) and 15′ of centrifugation, the fraction of large size flakes reduced slightly, still flakes with size up to 10 μm can be observed. In contrast, when both sonication and centrifugation times are increased (240′ and 90′), the distribution function localizes at small sizes with the mean (expected) value of ~ 1 μm and width of distribution of ~ 1.4 μm (see Figure 5(d), table 1).

Table 1. Parameters of flakes size distribution versus time of centrifugation of solution (DMF) produced with long time of sonication

centrifugation time (min)	L_C (μm)	$<L>$ (μm)	w (μm)
without	0.9	3.2	10.8
15	0.7	2.1	5.4
90	0.6	1.0	1.4

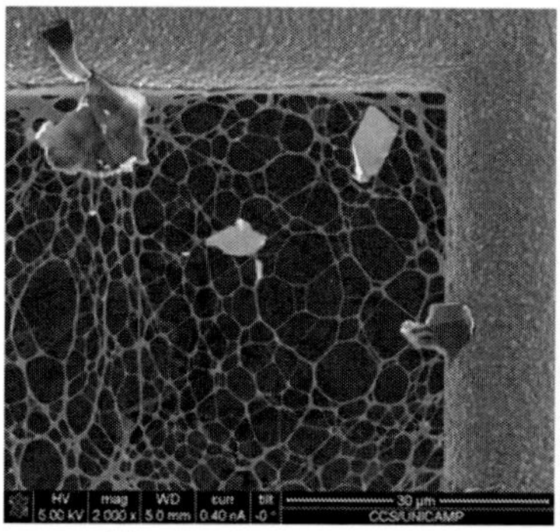

(a)

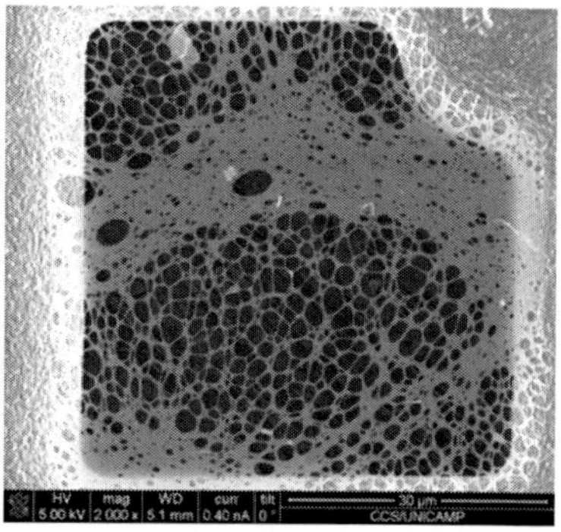

(b)

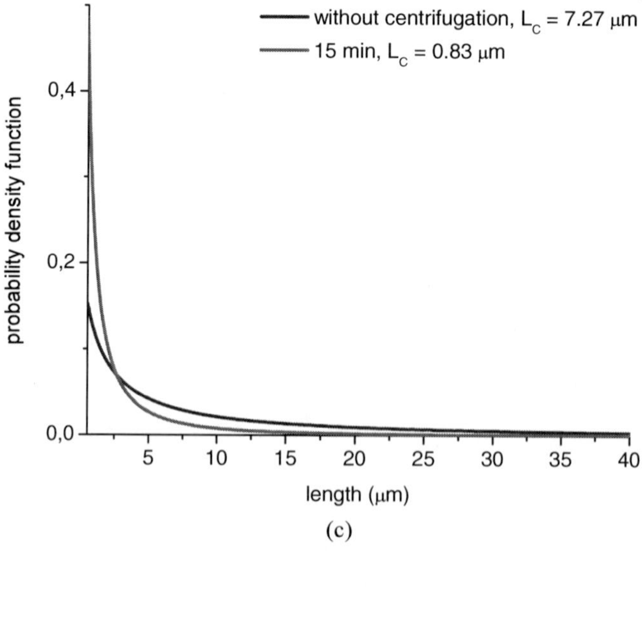

(c)

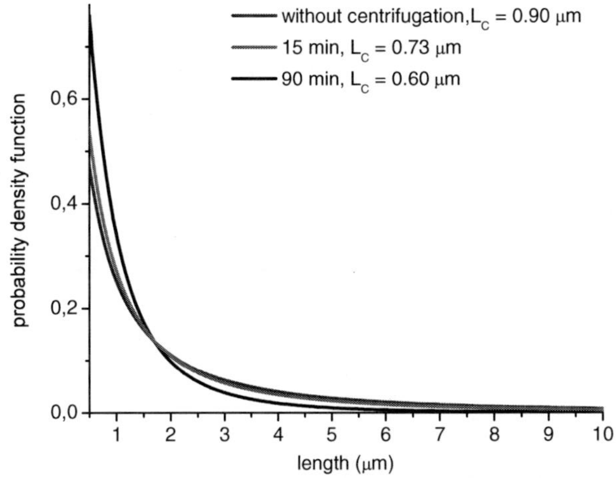

(d)

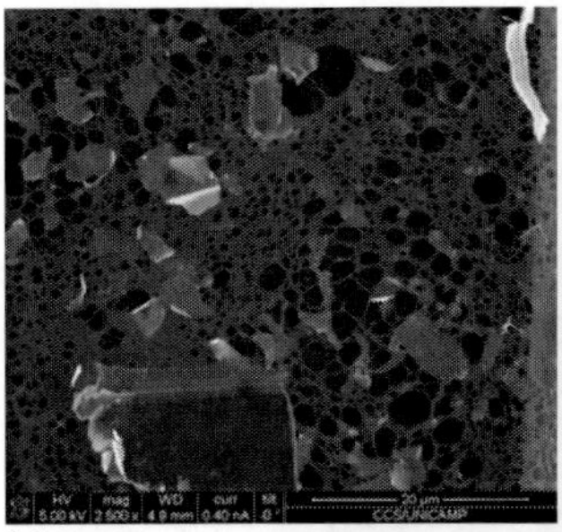

(e)

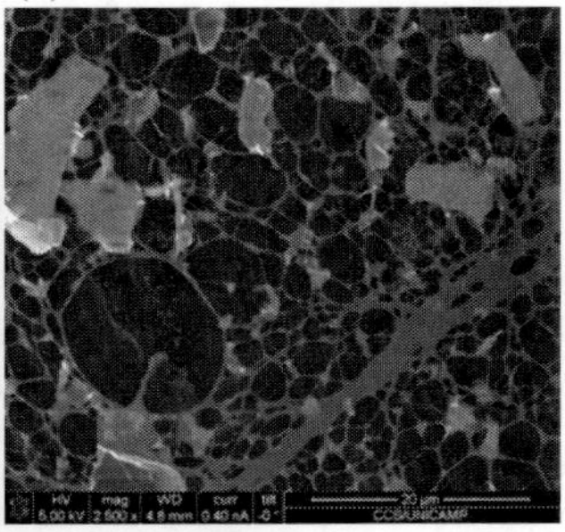

(f)

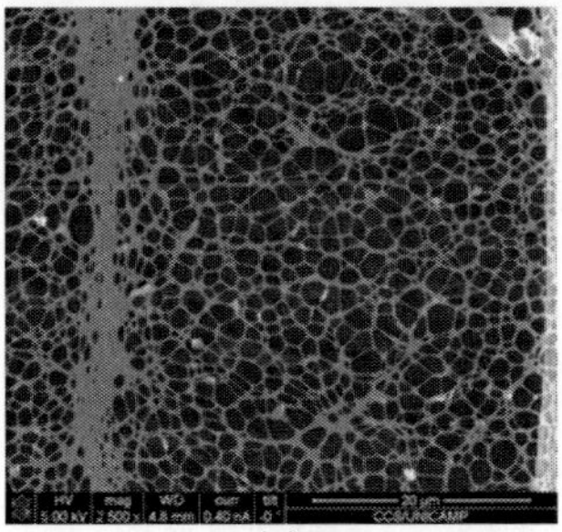

(g)

Figure 5. (a), (b) – SEM images of graphene's flakes deposited from suspension in DMF (time of sonication 2 min) as obtained and centrifuged with 800rpm during 15 min respectively, (c) – fitted curves of flakes size distribution as obtained and centrifuged with 800rpm during 15 min (time of sonication 2 min); (d) - fitted curves of flakes size distribution as obtained, centrifuged with 800rpm during 15 min and 90 min (time of sonication 240 min), (e), (f), (g) - SEM images of graphenes flakes deposited from suspension in DMF (time of sonication 240 min) as obtained, centrifuged with 800rpm during 15 min and 90 min respectively.

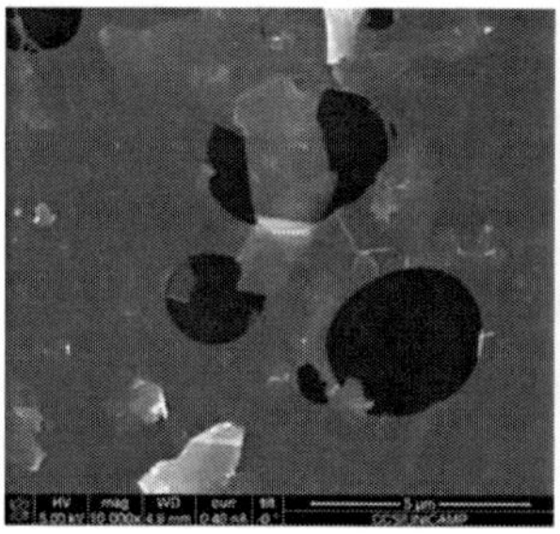

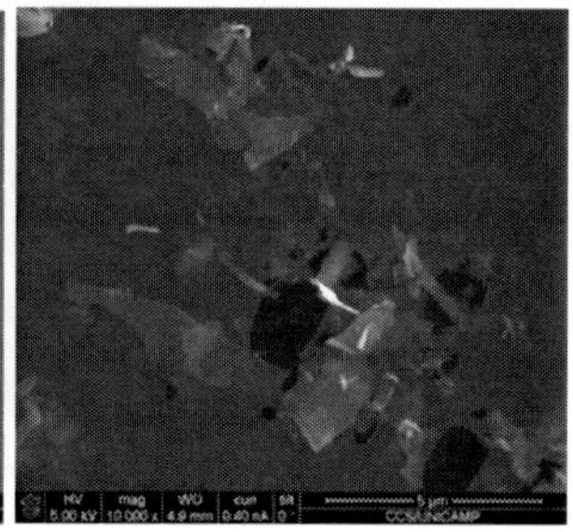

(a) (b)

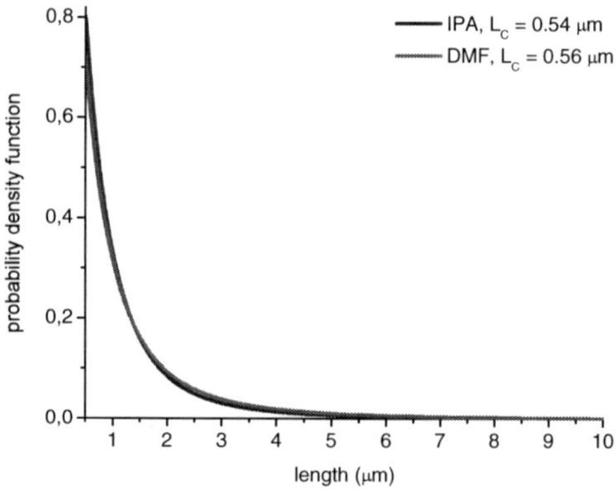

(c)

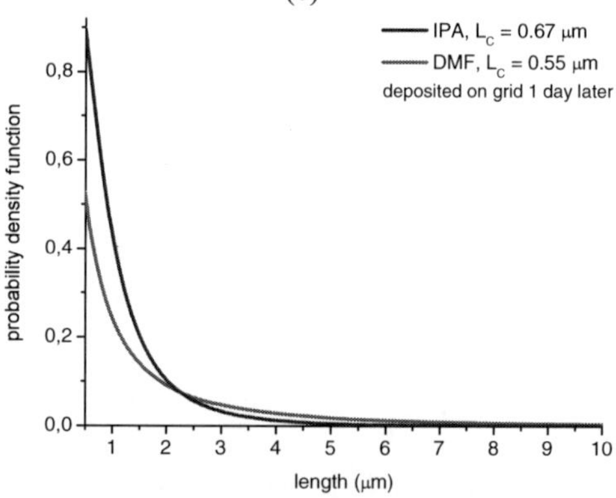

(d)

Figure 6. (a), (b) - SEM images of graphene's flakes deposited from suspension in DMF and IPA respectively; (c) - fitted curve of flakes size distribution for different solvents DMF and IPA, (d) – fitted curves of flakes size distribution for different solvents DMF and IPA deposited on grids after 1 day of fabrication. Time of sonication was 60 min, centrifugation with 800rpm during 15 min.

Analysis of the effect of solvent on the size distribution was done for DMF and IPA. In both cases, the same times of sonication and centrifugation were utilized, 60′ and 15′at 800 rpm, respectively. The images of lakes on TEM grids are shown in Figure 6 (a, b) and the fitted plot of size distribution is shown in Figure 6c.

From comparison of images and statistics results, it can be concluded that delamination of graphite occurs quite similarly for both solvents, no significant differences can be observed. The analysis was repeated after 1 day of letting suspensions in rest, revealing the difference in sedimentation of graphene flakes in these two solvents.

In DMF, the fraction of flakes with lateral sizes smaller than 2 μm reduced considerably, in contrast to IPA. This can be attributed to different dynamic viscosity of IPA and DMF, for IPA this value at 20°C is 2.5 higher than for DMF (2.43 mPa.s and 0.92 mPa.s, respectively). Considering that the viscous force is proportional to the contact area (flake size), the smaller flakes should deposit faster in less viscous media (DMF).

One of important factors in the graphite delamination is the solvent temperature. The process was carried out in IPA, with 120′ of sonication at 15°C e 40°C followed by centrifugation at 800 rpm for 60′. The samples obtained by deposition of suspension on TEM grids are shown in Figure 7, a, b and the size statistics is presented in Figure 7c and Table 2.

For the lower temperature, the distribution is more uniform and higher mean size, while for 40°C the curve is localized at smaller size, ~0.9 μm. This can be attributed to increased frequency of cavitation events (bubbles formation) and thus formation of smaller and more numerous graphene flakes, occurring at higher temperature.

Table 2. Parameters of flakes size distribution versus temperature of solvent (IPA)

temperature (°C)	L_C (μm)	$\langle L \rangle$ (μm)	w (μm)
15	0.81	1.8	3.5
40	0.57	0.9	1.1

Raman analysis was performed for flakes deposited on the TEM-grids to confirm high quality of graphene, with narrow full width at half maximum values, usually near 16-18 cm^{-1} (Figure 8(a, b)). The D/G band ratio that is widely used to evaluate the quality of the flakes, was found to vary from near

0 to 0.2 (versus time of sonication), a result which is consistent with small defect-free flakes.

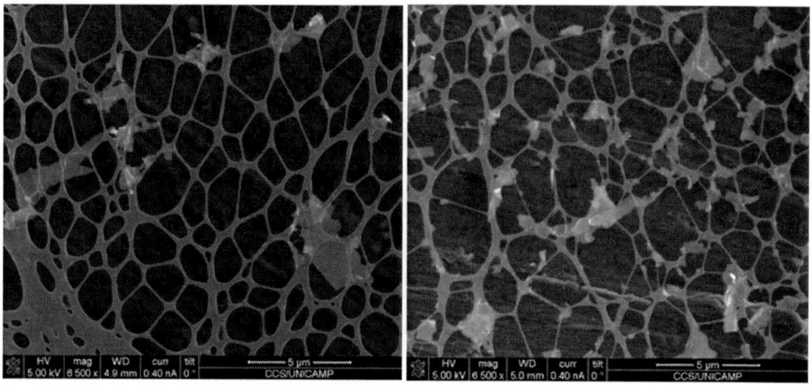

(a) (b)

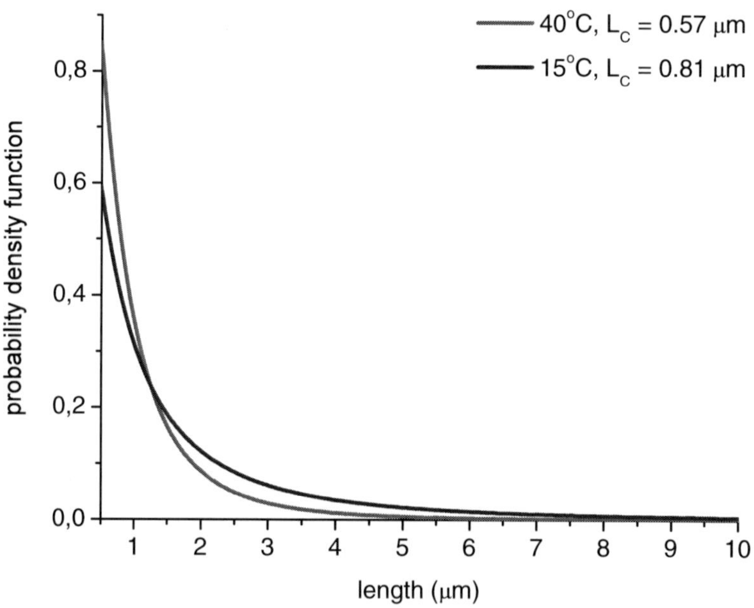

(c)

Figure 7. (a), (b) - SEM images of graphene's flakes deposited from suspension in IPA prepared with 15°C and 40 °C respectively, (c) - fitted curves of flak's size distribution versus different temperature. Time of sonication was 120 min, centrifugation with 800rpm during 60 min.

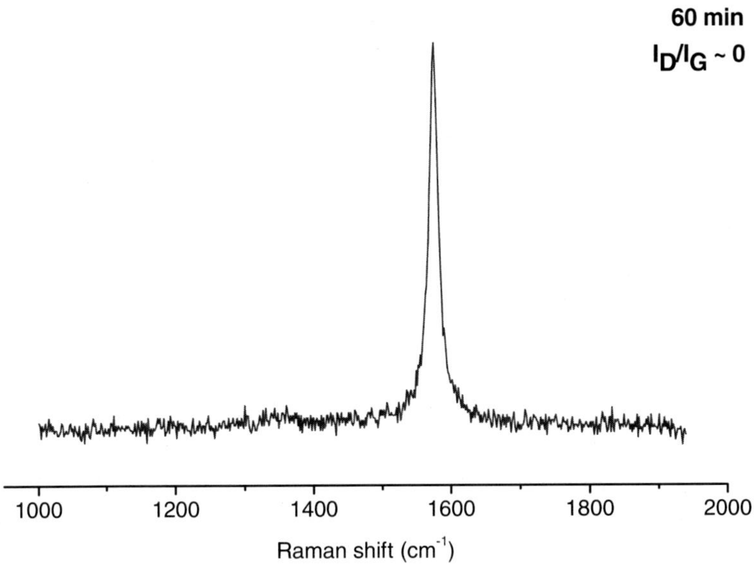

(a)

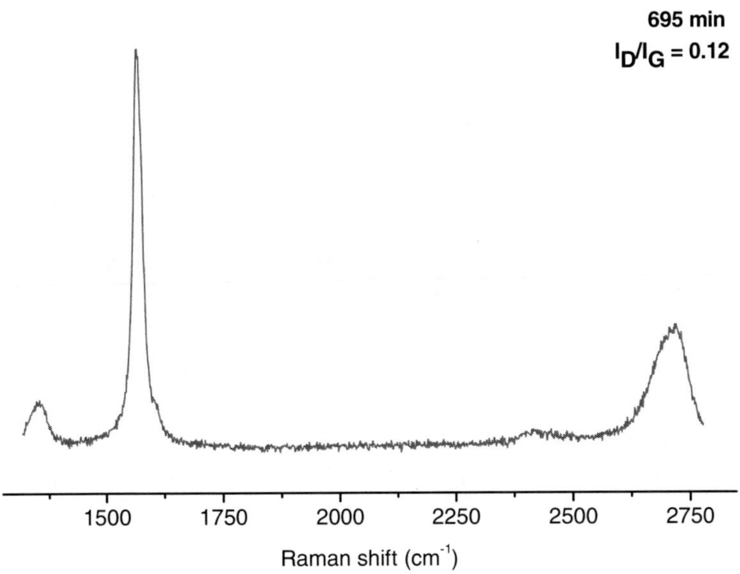

(b)

Figure 8. Raman spectra of graphene flakes (a) – at 60 min of sonication, (b) – at 695 min of sonication.

Conclusion

In the present work, a simple method to produce suspensions of high quality micron-size graphene flakes (multi-layer crystalline graphite or graphene). It was shown that lateral size distribution of multi-layer graphene flakes follows the log-normal distribution. The factors affecting the size distribution were analyzed. By changing the process conditions (sonication and centrifugation times) it was possible to control the mean sizes of graphene flakes.

Acknowledgments

The authors acknowledge financial support from INCT NAMITEC, CNPq and FAPESP (Brazil).

References

[1] K.S. Novoselov, A.K.Geim, S.V. Morozov, D. Jiang,Y. Zhang, S.V. Dubonos, I.V. Grigorieva, A.A. Firsov. *Science.* 306, (2004) 666-669.
[2] A.K. Geim, K.S. Novoselov. *Nat. Materials*. 6(3), (2007), 183.
[3] K.S. Novoselov, A.K.Geim, S.V. Morozov, D. Jiang, M.I. Katsnelson, I.V. Grigorieva, S.V. Dubonos, A.A. Firsov. *Nature.* 438, (2005) 197-200.
[4] M.H. Gass, U. Bangert, A.L. Bleloch, P. Wang, R.R. Nair, A.K. Geim, *Nat Nanotechnology*. 3, (2008). 676.
[5] Y. Kopelevich , P. Esquinazi , *Adv. Mater.* 19, (2007) 4559-4563.
[6] D. Graf, F. Molitor, K. Ensslin, C. Stampfer, A. Jungen, C. Hierold, L. Wirtz, *Nano Lett*. 7(2), 2007, 238-242.
[7] B.Z. Jang, A. Zhamu, *J. Mater. Sci.* 43, (2008) 5092-5101.
[8] S.Stankovich, D.A. Dikin, G.H.B. Dommet, K.M. Kohlhaas, E.J. Zimney, E.A. Stach, R.D. Piner, S.T. Nguyen, R.S. Ruoff, *Nature.* 442, (2006) 282-286.
[9] X. Liang, Z. Fu, S.Y. Chou, *Nano Lett.* 7, (2007) 3840.
[10] H. Yang, J. Heo, S. Park, H.J. Song, D.H. Seo, K.E. Byun, P. Kim, I. Yoo, H.J. Chung, K. Kim, *Science*. 336, (2012) 1140-1443.

[11] G. Konstantatos, M. Badioli, L. Gaudreau, J. Osmond, M. Bernechea, F. Pelayo Garcia de Arquer, F. Gatti, F.H.L. Koppens, Nat. Nanotechnology. 7, (2012) 363-368.

[12] P. Joo, B.J. Kim, E. K. Jeon, J.H Cho, B.S. Kim, *Chem. Commun.* 48, (2012) 10978–10980.

[13] Reina, X.T. Jia, J. Ho, D. Nezich, H. Son, V. Bulovic, M.S. Dresselhaus, J. Kong, *Nano Lett.*9(1), (2009) 30–35.

[14] H.X. Zhang, P.X. Feng, *Carbon.* 48, (2010) 359–364.

[15] Z.S. Wu, W.C. Ren, L.B. Gao, B.L. Liu, C.B. Jiang, H.M. Cheng, *Carbon*. 47, (2009) 493–499.

[16] G.H. Chen, W. Weng, D. Wu, C. Wu, J. Lu, P. Wang, X. Chen , *Carbon*. 42,(2004) 753-759.

[17] X. Li, X. Wang, L. Zhang, S. Lee, H. Dai, *Science*. 319, (2008) 1229-1232.

[18] Y. Hernandez, V. Nicolosi, M. Lotya, F.M. Blighe, Z. Sun, S. De, I.T. McGovern, B. Holland, M. Byrne, Y.K. Gun'ko, J.J. Boland, P. Niraj, G. Duesberg, S. Krishnamurthy, R. Goodhue, J. Hutchiso, V. Scardaci, A.C. Ferrari and J.N. Coleman. *Nat. Nanotechnology*. 3, (2008) 563-568.

[19] A.A. Green, M.C. Hersam. *Nano Lett*. 9(12), (2009) 4031-4036.

[20] M. Lotya, Y. Hernandez, P.J. King, R.J. Smith, V. Nicolosi, L.S. Karlsson, F.M. Blighe, S. De, Z. Wang, I.T. McGovern, G.S. Duesberg, J.N. Coleman. *J. Am. Chem. Soc*. 131, (2009) 3611-3620.

[21] M. Lotya, P.J. King, U, Khan, S. De, J.N. Coleman. *ACS Nano*. 4, (2010) 3155-3162.

[22] U. Khan, A. O'Neill, M. Lotya, S. De, J.N. Coleman. *Small*. 6, (2010) 864–871.

[23] U. Kahn, A. O'Neill, H. Porwal, P. May, K. Nawaz, J.N. Coleman. *Langmuir*. 27, (2011) 9077-9082.

[24] U. Kahn, A. O'Neill, H. Porwal, P. May, K. Nawaz, J.N. Coleman. *Carbon*. 50, (2012) 470-475.

[25] M. Yi, Z. Shen, S. Ma, X. Zhang. *J. Nanopart. Res*. 14, (2012) 1003.

[26] J. N. Coleman. *Adv. Funct. Mater*. 19, (2009) 3680-3695.

[27] Y. Hernandez, M. Lotya, D. Rickard, S.D. Bergin, J.N. Coleman. *Langmuir*. 26, (2010) 3208-3213.

[28] A. O'Niell, U. Khan, P.N. Nirmalraj, J. Boland, J.N. Coleman. *J. Phys. Chem. C*. 115, (2011) 5422-5428.

[29] C. Panayiotou. *Phys. Chem. Chem. Phys*. 14, (2012) 3882-3908.

[30] S. E. Skrabalak. *Phys. Chem. Chem. Phys*. 11, (2009) 4930-4942.

[31] G. Cravotto, P. Cintas. *Chem. Eur. J*. 16, (2010) 5246 – 5259.

[32] K. S. Suslick, D.J. Flannigan. *Ann. Rev. Phys. Chem.* 59, (2008) 659-683.

[33] T. J. Mason. *Practical sonochemistry: user's guide to applications in chemistry and chemical engineering*, Ellis Horwood series in organic chemistry, London, 1991, pp. 27-28.

[34] A. N. Kolmogorov. *The proceedings of the USSR Academy of Sciences.* A. E. Fersman; Mathematics; Moscow, 1941; 31(2), 99-101.

[35] H. Chernoff, E.L. Lehmann. *The Annals of Mathematical Statistics.* 25 (3), (1954) 579–586.

[36] R. L. Plackett. *International Statistical Review*. 51 (1), (1983) 59–72.

[37] L. N. Bol'shov, Y.V. Smirnov. *Tables of Mathematical Statistics*; Moscow, 1983; 166-169.

In: Smart Nanoobjects
Editor: Kirill Levine
ISBN: 978-1-62808-741-3

Chapter 9

SYNTHESIS AND CHARACTERIZATION OF SOME TRANSITION METAL COMPLEXES OF 1-(BENZOTHIAZOL-2-ACETYL)-4-PHENYLTHIOSEMICARBAZIDE

***Ayman K. El-Sawaf*[1,2*], *Fathy A. El-Saied*[2] and *Ahmed M. A. El-Seidy*[3]**

[1]College of Science and Humanity Studies, Salman bin Abdulaziz University, KSA

[2]Chemistry Department, Faculty of Science, El-Minufiya University, El-Minufiya, Egypt

[3]Inorganic Chemistry Department, National Research Centre, Dokki, Giza, Egypt

ABSTRACT

A novel ligand H_2L (**1**), 1-(benzothiazol-2-acetyl)-4-phenylthiosemicarbazide was synthesized by the reaction of benzothiazole-2-ylacetohydrazide with phenyl isothiocyanate. This ligand was used to prepare new transition metal complexes of Fe(III), Co(II), Ni(II), Cu(II), and Zn(II). Characterization of the H_2L (**1**) ligand and its complexes is also reported. Mass spectra and NMR assignments for the

* E-mail: elsawaf2008@yahoo.com; Fax: 00966015453658.

ligand were carried out. Electronic and magnetic moments of the complexes indicate that the geometries of the metal centers are either distorted octahedral, octahedral, tetrahedral or square planar. The structures are consistent with the IR, UV-VIS as well as conductivity measurements.

Keywords: Metal complexes, thiosemicarbazide, benzothiazole, isothiocyanate, Characterization

1. INTRODUCTION

Thiosemicarbazones belong to a group of thiourea derivatives and have emerged as an important class of sulfur donor ligands particularly for transition metal ions. For long time, semi- and thiosemicarbazones has been a subject of interest to researchers of different profiles. In view of the fact that these compounds form with many metals complexes of diverse chemical, physical and structural characteristics, they are of special interest to coordination chemists [1–3].

Metal complexes with these Schiff bases have the ability to reversibly bind oxygen in epoxidation reactions [4], catalytic activity in hydrogenation of olefins [5] and photochromic properties [6]. Moreover, the chemistry of metal complexes of Schiff base ligands having nitrogen, oxygen and sulfur at their donor sites has been extensively studied since the late 19th century [7-12]. Also, these complexes have numerous applications, such as, in the treatment of cancer [13], as antibactericide agents [14], as antiviral agents [15], as fungicide agents [16] and for other biological properties [17]. All these advantages make Schiff bases very good candidates in the effort to synthesize metal complexes of interest in bioinorganic chemistry, catalysis, encapsulation, transport and separation processes, magnetochemistry [18] and heterogeneous and homogeneous catalysis for oxidation and polymerization of organic compounds [19, 20].

Copper complexes with bis(thiosemicarbazone) ligands synthesized in the 1950s [21], have anticancer chemotherapeutic [22] and superoxide dismutase-like activity [23] and in the radiolabelled form have been used as positron emission agents for blood perfusion [24] and most recently tissue hypoxia [25]. The use of small magnetic centers has gained much interest because of the magnetic materials based on metal-cyano bridges [26]. Based on the aforementioned properties for Schiff bases and their complexes, we report herein the synthesis and spectroscopic studies of the ligand H_2L, 1-

(benzothiazol-2-acetyl)-4-phenylthiosemicarbazide and its Fe(III), Co(II), Ni(II), Cu(II) and Zn(II) complexes.

2. EXPERIMENTAL

2.1. Material

All the reagents employed for the preparation of the ligand and its complexes were of the best grade available and used without further purification. Solvents employed were dried and distilled before use by standard methods.

2.2. Synthesis of the Schiff Base Ligand H_2L

The Schiff base ligand H_2L was prepared in two steps. The reactions steps are illustrated in Scheme 1.

2.2.1. Synthesis of Benzothiazol-2-ylacetohydrazide

A hot (75° C) solution of 2-aminothiophenol (126 mg, 1 mmol) in ethanol (20 ml) was added to a hot (75° C) stirred solution containing ethylcyanoacetate (128 mg, 1 mmol) and few drops of pyridine in ethanol (15 ml) then reflux for 4 h. 7 mL of hydrazine hydrate was then added dropwise to the reaction mixture over a period of 5-8 min and continue refluxing for another hour. After coaling the precipitate was filtered off, crystallized from ethanol and dried under vacuum over anhydrous $CaCl_2$ (mp = 155-157° C).

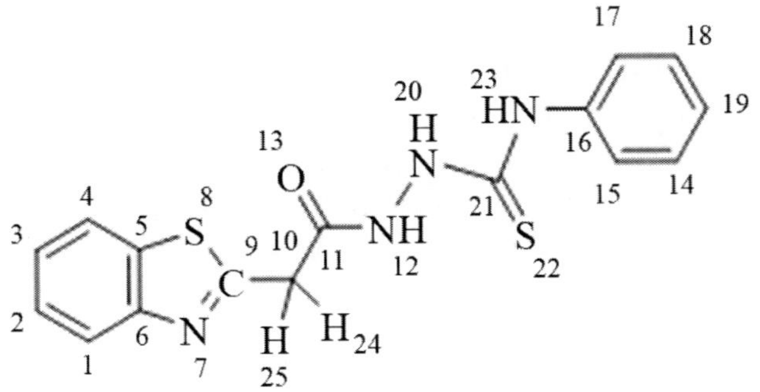

Chart 1. Numbering scheme for ligand H_2L.

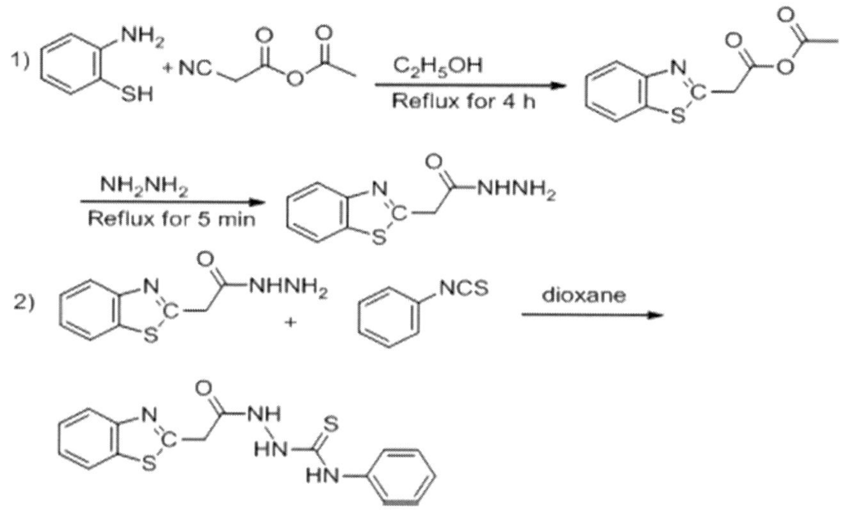

Scheme 1. Schematic representation for the formation of the ligand H_2L.

2.2.2. Synthesis of 1-(benzothiazol-2-acetyl)-4-phenylthiosemicarbazide

A hot (75° C) solution of benzothiazol-2-ylacetohydrazide (207 mg, 1 mmol) in dry dioxane (15 mL) was added to a hot (75° C) solution containing phenyl isothiocyanate (135 mg, 1 mmol) and 3 drops of triethylamine in dioxane (15 mL) and reflux for 15 min. After coaling the precipitate was filtered off, crystallized from dioxane and dried under vacuum over anhydrous $CaCl_2$ (Scheme 1). (mp = 216-218° C, yield = 80%).

2.2.3. Synthesis of the Metal Complexes

The metal complexes of the ligand H_2L were prepared by mixing a hot (60^0 C) methanolic solution of the metal salts: $FeCl_3.6H_2O$, $CoCl_2.6H_2O$, $CoBr_2.6H_2O$, $Co(OAc)_2.4H_2O$, $Co(NO_3)_2.6H_2O$, Co(II) perchlorate, $NiCl_2.6H_2O$, $Ni(C_2H_3O_2)_2.4H_2O$, $Ni(NO_3)_2.6H_2O$, $CuCl_2.2H_2O$, $CuBr_2$, $Cu(CH_3COO)_2.H_2O$, $Cu(NO_3)_2.3H_2O$, $ZnCl_2$ with the required amount of a hot (75^0 C) ethanolic solution of the ligand to form 1:1 or 1:2 L/M (ligand /metal).

The reaction mixture was then refluxed for a time depending on the transition metal salt used. The precipitates formed were filtered, washed with ethanol, then with diethyl ether and dried under vacuum over anhydrous $CaCl_2$.

2.4. Physical Measurements

The ligand and its metal complexes were analyzed for C, H, N and M contents at the Microanalytical Laboratory, Faculty of Science, Cairo University, Egypt. Analytical and physical data of the ligand H_2L and its metal complexes are reported in Table 1.

The metal ion contents of the complexes were also determined [27] by the previously reported methods [28]. IR spectra of the ligands and their metal complexes were measured using KBr discs with a Jasco FT/IR 300E Fourier transform infrared spectrophotometer covering the range 400–4000 cm^{-1} and in the 500–100 cm^{-1} region using polyethylene-sandwiched Nujol mulls on a Perkin Elmer FT-IR 1650 spectrophotometer. ^1H NMR spectra was obtained on 300 MHz VARIAN NMR spectrometers. Chemical shifts (ppm) are reported relative to TMS. The mass spectra were run at 70 eV with HP MODEL: MS. 5988A. The electronic spectra of the ligands and their complexes were obtained in Nujol mulls and in saturated DMSO solutions using a Shimadzu UV–240 UV–Vis recording spectrophotometer. Molar conductivities of the metal complexes in DMSO (10^{-3} M) were measured using a dip cell and a Bibby conductimeter MC1 at room temperature. The resistance measured in ohms and the molar conductivities were calculated according to the equation: $\Lambda = V \times K \times Mw/g \times \Omega$, where Λ, molar conductivity (ohm^{-1} cm^2 mol^{-1}); V, volume of the complex solution (mL); K, cell constant 0.92 cm^{-1}; Mw, molecular weight of the complex; g, weight of the complex; and Ω, resistance measured in ohms. Magnetic moments at 298 K were determined using the Gouy method with Hg $[Co(SCN)_4]$ as calibrant.

3. Results and Discussion

3.1. Mass Spectra of the Ligand and Complexes

The mass spectra of the Schiff base ligand H_2Lsupported the suggested structure of the ligand. It reveals molecular ion peak m/e at 342, consistent with the molecular weight of the ligand. Moreover the fragmentation pattern splits a parent ion peak at m/e = 207, which corresponding to $C_9H_9N_3SO$, while the fragment at m/e = 135, 74, 58 and 77 corresponding to C_7H_5NS, $C_2H_6N_2O$, C_2H_4NO and C_6H_5 respectively.

The mass spectra of the complexes did not show the parent peaks corresponding to their Molecular weights. This finding may be attributed to that the working temperature inside the mass spectrometer is relatively high; so that the mass spectra of the complexes are in fact the spectra of the decomposition products at working temperature.

3.2. ^{1}H NMR Spectrum of the Ligand

The ^{1}H NMR spectrum of the ligand in d-chloroform, which would produce more information concerning intramolecular hydrogen bonding [29], was not possible due to their low solubility, so it has been recorded as d6-DMSO solution.

The resonance for the amido N(23)H attached to the phenyl group is located at 9.8 ppm, indicating that hydrogen bonding with d6 –DMSO does not occur, in agreement with previous results [30]. The H signal due to the hydrazido group N(20) H occurs at $\delta = 10.5$ ppm indicating the involvement of this hydrogen thorugh intramolecular hydrogen bonding with the carbonyl oxygen of –C-NH group. The other hydrazido group N(12)H appears at $\delta = 9.6$ ppm. A singlet at $\delta = 4.2$ ppm and the multiplet at 8.13 – 7.43 ppm are arrributed to the protons of CH_2 and aryl groups, respectively [31].

3.3. Infrared Spectra of the Complexes

IR spectra of the complexes were recorded to confirm their structures. The most characteristic vibrational frequencies and their tentative assignments for the ligand H_2L and its transition metal complexes are listed in Table 2. The assignments were made by comparison with the vibrational frequencies of the free ligand.

The two strong bands at 1685 and 780 cm^{-1} were assigned to ν(C=O) and ν(C=S) vibrations. The absence of any band above 3400 cm^{-1} or in the region 2550-2800 cm^{-1} region suggests the absence of any hydroxide or thiol tautomer in the solid state [32, 33]. The absence of two broad bands at ca. 2120 and 1820 cm^{-1} which are due to NH······O stretching and bending vibrations [34] may suggest the absence of interamolecular hydrogen bonding in the solid state. The three bands at 3320, 3250 and 3170 cm^{-1} were assigned to ν(N^4H), ν(N^2H) and ν(N^1H) respectively, while the ν(N-N) vibration is observed at 920 cm^{-1} [35] as medium sharp band. The bands at 1500, 1420 and 1260 cm^{-1} may be due to ν(-N-C=S) [36]. These bands are assigned as coupled modes consisting principally of ν(NH) and ν(CN).

The IR spectra of the complexes $[Cu(HL)_2(H_2O)_2]H_2O$ and $[Zn(HL)_2].3H_2O$ showed that the ligand behaves as mononegative bidentate and coordinating via the N^2H group and the enolic carbonyl oxygen with the displacement of the hydrogen atom from the latter group, forming a five-membered ring including the metal atom [scheme 2]. This mode of chelation is supported by the following evidence: (i) the disappearance of ν(C=O) with the appearance of new bands in the 1600-1585 cm^{-1} and 1200-1170 cm^{-1} regions assigned to ν(O-C=N) [51] and ν(C-O) [37] respectively; (ii) ν(N-N) shifts to a higher wave number ca. 1060 cm^{-1}; (iii) the appearance of new bands in the regions 460-430 cm^{-1} and 390-380 cm^{-1} assigned to ν(M-N) [38] and ν(M-O) [39], respectively.

Some complexes included water molecules. The broad bands in the 3600-3400 cm^{-1} region are due to coordinated water or water of crystallization. The bands for water or crystallization are different from those of coordinated water, the latter has a band in the 600-400 cm^{-1} region, but the absence of this bands in the spectrum or the Zinc(II) complex, indicates the hydrated water rather than coordinated ones. The presence of water molecules within the coordination sphere in the hydrated $[Cu(HL)_2(H_2O)_2]H_2O$ complex is supported by the presence of bands at 3480 cm^{-1}, 1610 cm^{-1}, 950 cm^{-1} and 630 cm^{-1} due to OH stretching, HOH deformation, H_2O rocking and H_2O wagging, respectively [40]. The absence of coordinated water molecules in complexes $[Co(H_2L)_2(NO_3)_2].3H_2O$, $[Ni(H_2L)_2(NO_3)_2].3H_2O$ and $[Cu(H_2L)_2(NO_3)_2]$. $2H_2O$ was confirmed from the absence of the rocking, twisting and wagging vibrational modes which are normally activated at 970-930 cm^{-1} and 660-600 cm^{-1} [41], as well as the presence of weak and broad band at 3450 cm^{-1} indicating that the water in all of these complexes is lattice [42] rather than coordinated.

For the complexes $[Co(HL)(OAc)]H_2O$ and $[Ni(HL)_2]H_2O$ the ligand behaves as mononegative bidentate coordinating via the N^1H group and the thiol sulfur atom with loss of the proton from the later group, This mode of complexation, scheme 2, confirmed by the following: (i) the disappearance of ν(C=S) with the appearance of new bands due to ν(S-C=N) [40] and ν(C-S) [43] at about 700 cm^{-1} and 360 cm^{-1} respectively. As expected, greater decrease in the thioamide (IV) band occurs for the anionic form of the ligand due to C=S formally becoming a single bond (C-S) [44]; (ii) ν(C=O) Remains more or less at the same position; (iii) the appearance of new bands at 430 cm^{-1} and 360 cm^{-1} assignable to ν(M-N) [45] and ν(M-S) [46] respectively.

Table 1. Analytical and physical data of the ligand H_2L and its metal complexes

No.	Ligand/ Complexes	Colour	FW	Yield (%)	Analysis (%) / Found (calcd)				Molar Conductance Λ_m ($\Omega^{-1}cm^2mol^{-1}$)
					C	H	N	M	
1	H_2l	Yellow	342.4	80	56.0(56.1)	4.2(4.1)	16.5(16.4)	-	-
2	$Fe(H_2L)(HL)Cl_2$	brown	810.6	78	47.9(47.4)	3.5(3.4)	14.0(13.8)	6.6(6.9)	8
3	$[Co(H_2L)_2Cl_2]H_2O$	brown	832.7	70	46.3(46.2)	3.8(3.6)	13.7(13.5)	6.9(7.1)	11
4	$[Co(H_2L)Br_2]H_2O$	Red	579.20	72	33.1(33.2)	3.0(2.8)	9.8(9.7)	10.0(10.2)	7
5	$[Co(HL)(OOCCH_3)]H_2O$	brown	477.4	74	45.1(45.3)	4.0(3.8)	11.8(11.7)	12.0(12.3)	13
6	$[Co(H_2L)_2(NO_3)_2].3H_2O$	brown	921.9	67	41.4(41.7)	3.8(3.7)	15.3(15.2)	6.3(6.4)	15
7	$[Co(H_2L)(HL)]ClO_4$	brown	842.3	66	45.4(45.6)	3.3(3.2)	13.3(13.3)	6.9(7.00)	89
8	$[Ni(HL)_2]H_2O$	green	741.6	82	51.8(51.8)	3.7(3.5)	15.2(15.1)	7.8(7.9)	10
9	$Ni(H_2L)_2(OC(O)CH_3)_2$	brown	861.7	86	50.1(50.2)	4.2(4.0)	13.1(13.0)	7.0(6.8)	9
10	$[Ni(H_2L)_2(NO_3)_2].3H_2O$	brown	921.6	79	41.5(41.7)	3.9(3.7)	15.3(15.2)	6.4(6.4)	10
11	$[Cu(H_2L)_2Cl_2]H_2O$	brown	855.4	81	44.8(44.9)	4.0(3.8)	13.3(13.10)	7.2(7.4)	12
12	$[Cu(H_2L)_2Br_2]H_2O$	brown	926.3	78	41.4(41.5)	3.6(3.3)	12.2(12.1)	6.7(6.9)	13
13	$[Cu(HL)_2(H_2O)_2]H_2O$	brown	800.5	89	48.0(48.0)	4.1(4.0)	14.1(14.0)	7.8(7.9)	6
14	$[Cu(H_2L)_2(NO_3)_2].2H_2O$	green	908.5	84	42.1(42.3)	3.8(3.6)	15.5(15.4)	6.9(7.0)	14
15	$[Zn(HL)_2].3H_2O$	beig	802.3	69	47.8(47.9)	4.1(4.0)	14.2(14.0)	8.0(8.2)	7

Table 2. IR frequencies of the bands (cm^{-1}) of ligand H_2L and its complexes and their assignments

No.	Ligand/ Complexes	ν(N^4H)	ν(N^4H)	ν(N^4H)	ν(C=O)	ν(C=S)	ν(N-N)	ν(M-N)	ν(M-O)	ν(M-S)
1	H_2L	3320m	3250w	3170m	1685s	780s	1120s	-	-	-
2	$Fe(H_2L)(HL)Cl_2$	3310m	3220w	3160w	1630m	740m	1140m	460m	420m	-
3	$[Co(H_2L)_2Cl_2]H_2O$	3300m	3220w	3180m	1630m	750s	1135m	495m	425m	-
4	$[Co(H_2L)Br_2]H_2O$	3340w	3240w	3190w	1640w	765s	1151m	480w	430w	-
5	$[Co(HL)(OOCCH_3)]H_2O$	3200m	-	3180w	1660w	690m	1127m	430m	410w	365m
6	$[Co(H_2L)_2(NO_3)_2].3H_2O$	3340w	3250w	3160w	1595m	755m	1147m	430w	355w	-
7	$[Co(H_2L)(HL)]ClO_4$	3320m	-	3160m	1630m	760m	1137m	440w	340w	-
8	$[Ni(HL)_2]H_2O$	3300m	-	3180w	1640m	720m	1131m	430m	-	360m
9	$Ni(H_2L)_2(OC(O)CH_3)_2$	3290m	3220m	3150w	1645m	765s	1129w	440m	380m	-
10	$[Ni(H_2L)_2(NO_3)_2].3H_2O$	3300m	3230w	3180w	1635m	750s	1146m	425m	360m	-
11	$[Cu(H_2L)_2Cl_2]H_2O$	3290m	3210w	3140w	1640m	765s	1149m	430m	375m	-
12	$[Cu(H_2L)_2Br_2]H_2O$	3295m	3210m	3160w	1650m	725m	1132w	440w	-	370m
13	$[Cu(HL)_2(H_2O)_2]H_2O$	3305m	-	3140w	-	760s	1136m	430m	380m	-
14	$[Cu(H_2L)_2(NO_3)_2].2H_2O$	3295m	3220w	3150w	1620m	755s	1140m	440m	370w	-
15	$[Zn(HL)_2].3H_2O$	3310s	-	3140w	-	760s	1137m	460m	390m	-

Table 3. The electronic absorption spectral bands (nm) and magnetic moment (B.M.) for the ligand H_2L and its complexes

No.	Ligand/ Complexes	mode	π–π* , n–π* and CT bands				d–d bands			μ_{eff} in BM
1	H_2L	solid	289.9	341.9	388.5	-	-	-	-	-
		In DMF	290.0	340.0	350.0	450.0	-	-	-	
2	$Fe(H_2L)(HL)Cl_2$	solid	293.1	332.0	351.9	406.2	487.8	648.9	-	5.7
		In DMF	290.3	331.2	349.3	432.0	520.0	550.1	-	
3	$[Co(H_2L)_2Cl_2]H_2O$	solid	293.7	321.9	336.1	399.4	486.1	649.4	-	4.1
		In DMF	290.8	321.0	345.5	414.1	520.0	559.9		
4	$[Co(H_2L)\ Br_2]H_2O$	solid	296.0	334.0	363.9	397.6	505.1	666.7	-	2.0
		In DMF	291.4	331.5	349.4	430.3	500.0	570.1	719.9	
5	$[Co(HL)(OOCCH_3)]H_2O$	solid	290.5	323.9	351.2	400.3	497.5	665.8	-	2.1
		In DMF	290.6	330.7	344.7	417.7	590.0	-	-	
6	$[Co(H_2L)_2(NO_3)_2].3H_2O$	solid	294.6	325.3	349.2	400.6	469.5	625.0	-	3.8
		In DMF	290.0	331.7	347.7	414.9	595.9	-	-	
7	$[Co(H_2L)(HL)]ClO_4$	solid	295.5	335.0	363.8	405.2	498.0	668.0	-	2.2
		In DMF	290.1	325.3	398.2	-	600.6	-	-	
8	$[Ni(HL)_2]H_2O$	solid	293.3	339.3	384.5	400.6	630.1	833.3	-	2.8
		In DMF	292.1	330.7	343.1	-	566.6	600.2	-	
9	$Ni(H_2L)_2(OC(O)CH_3)_2$	solid	294.6	339.9	377.6	397.1	649.8	-	-	3.0
		In DMF	292.4	335.8	347.9	396.8	607.9	-	-	
10	$[Ni(H_2L)_2(NO_3)_2].3H_2O$	solid	294.0	331.7	370.2	393.9	626.2	-	-	2.5
		In DMF	290.8	333.9	358.8	412.0	595.9	-	-	
11	$[Cu(H_2L)_2Cl_2]H_2O$	solid	294.4	334.0	357.9	383.1	609.8	-	-	1.79
		In DMF	290.0	330.0	360.1	390.0	550.1	605.0	669.8	
12	$[Cu(H_2L)_2Br_2]H_2O$	solid	296.4	330.7	355.9	392.3	605.7	-	-	1.8
		In DMF	290.2	332.2	354.1	-	547.3	650.2	-	
13	$[Cu(HL)_2(H_2O)_2]H_2O$	solid	292.8	341.2	377.8	403.9	590.0	-	-	1.75
		In DMF	290.8	331.1	349.2	402.6	539.4	580.0		
14	$[Cu(H_2L)_2(NO_3)_2].2H_2O$	solid	291.8	331.8	389.3	412.5	599.9	-	-	1.77
		In DMF	290.2	493.8	359.2	396.5	536.5	637.8	-	
15	$[Zn(HL)_2].3H_2O$	solid	288.9	330.7	-	-	-	-	-	-
		In DMF	290.3	326.1	342.0	-	-	-	-	

Extensive IR spectral studies reported on metal aceto complexes [47] indicate that the acetate ligand may coordinate to a metal center in either a monodentatc, bidentate or bridging manner.

The $\nu_{asym.}(CO_2)$ and $\nu_{sym.}(CO_2)$ of the free acetate ions are at 1560 cm^{-1} and 1416 cm^{-1}, respectively. In monodcntatc coordination ν(C=O) is found at higher energy than $\nu_{asym.}(CO_2)$ and ν(C-O) is lower than $\nu_{sym.}(CO_2)$. As a result, the separation between the two ν(CO) bands is much larger in monodentate complexes than the free ion [48]. The opposite trend is observed in bidentate aceto coordination; the separation between ν(CO) is smaller than for the free ion as found for the complex $[Co(HL)(OOCCH_3)]H_2O$, $\nu_{asym.}(COO^-)$ and $\nu_{sym.}(COO^-)$ bands at 1570 cm^{-1} and 1450 cm^{-1}, respectively. For bridging acetate with both oxygens coordinated as in copper(Il) acetate, however, the two ν(CO) bands are close to the free ion values [49]. The aceto group in the complex $Ni(H_2L)_2(OC(O)CH_3)_2$ acts as a unidentate ligand, and this is supported by the appearance of two new bands at 1610 cm^{-1} and 1350 cm^{-1}, which may be attributed to $\nu_{asym.}(COO^-)$ and $\nu_{sym.}(COO^-)$, respectivel [49]. Further, the complex exhibits δ(COO) at 750 cm^{-1} which is considered diagnostic for unidentate acetates [50]. The infrared spectra for the rest of the complexes show that the ligand behaves as neutral bidentate, coordinating via the carbonyl oxygen (C=O) and the nitrogen atom of the N^2H group. The mode of complexation is supported by the shift of both ν(C=O) and $\nu(N^2H)$ to lower wavenumber with the appearance of new bands assignable to ν(M-O) and ν(M-N) as shown in table 2. The spectra of the complexes $[Co(H_2L)_2Cl_2]H_2O$ and $[Cu(H_2L)_2Cl_2]H_2O$ show two assignable, weak intensity ν (M-CI) bands at 320 cm^{-1} and 310 cm^{-1} indicating of a terminal Chloro ligands [51].

The spectra of the nitrato complexes, $[Co(H_2L)_2(NO_3)_2].3H_2O$, $[Ni(H_2L)_2(NO_3)_2].3H_2O$ and $[Cu(H_2L)_2(NO_3)_2].2H_2O$ showed a band near 290 cm^{-1} which can be assigned to ν(M-O), $\nu(M\text{-}ONO_2)$, by analogy with the bands at 250-300 cm^{-1} reported earlier for $Cu\text{-}ONO_2$ in metal complexes [52] containing nitrate as monodentate ion. Moreover, in the nitrato complexes two bands at 1290 cm^{-1} and 1400 cm^{-1} corresponding to ν_1 and ν_4 modes of nitrato groups with ion of 100 cm^{-1} indicate the presence of terminally bonded monodentate nitrate group. The infrared spectrum of the cobalt(Il) complex $[Co(H_2L)(HL)]ClO_4$ Showed a strong unsplit band (ν_3) at 1090 cm^{-1} and a strong band at 610 cm^{-1} (ν_4) which are indicative of tetrahedral symmetry for the uncoordinated perchlorate ion [53].

3.4. Molar Conductance Data

All metal complexes are stable in air and insoluble in common organic solvents but easy soluble in DMF and DMSO. The molar conductivities of the complexes in DMF (10^{-3} M) are listed in table 1. Only the complex $[Co(H_2L)(HL)]ClO_4$ behaves as a 1:1 electrolyte. The rest of the complexes show a non electrolyte nature [54].

3.5. Electronic Spectra and Magnetic Moments

The solid state spectrum of the thiosemicarbazide H_2L shows a broad band at 289.9 nm assignable to the phenyl ring $\pi \rightarrow \pi^*$ transitions. This band is shifted to longer wave length which is consistent with an increase in the degree of π-cloud conjugation [55]. The $n \rightarrow \pi^*$ transition associated with the carbonyl portion of the thiosemicarbazide moiety around 341.9 nm in the spectrum of the free ligand and is generally shifted to higher energy on complexation. *A* second $n \rightarrow \pi^*$ band originating from the thioamide portion of the thiosemicarbazide moiety is found at about 388.5 nm in the spectrum of H_2L. In the spectra of some complexes this latter band generally shifts to higher energies and sometimes merges with the $n \rightarrow \pi^*$ transition associated with the band due to the carbonyl function.

There are also charge-transfer transitions partially responsible for the intense colors of the complexes. Present at energies above 333 nm in the spectra of the metal complexes are S→M(II), M(III) charge transfer bands [56], as well as Br→M(II) [57] charge transfer bands. The Cl→ M(II) [58] and O→ M(II) [59] bands are generally found in the 333 nm region of the spectra and would be obscured by the stronger intraligand bands. This series of complexes has very broad bands due to the various $n \rightarrow \pi^*$ transition, making it difficult to make assignments.

In DMF solution of H_2L the intense band ($\log \varepsilon = 4.52$) is the $n \rightarrow \pi^*$ transition of the carbonyl portion while the shoulder at 450.0 nm is the $n \rightarrow \pi^*$ transition associated with the thione portion of the thiosemicarbazide moiety. These bands are shifted to higher energy in DMF compared to the solid state, presumably due to the intermolecular hydrogen bonding between N^2H and DMF. A peak associated with the phenyl $\pi \rightarrow \pi^*$ transition is observed at about 290.0 nm in the spectrum of the free thiosemicarbazide and is nearly unshifted in the spectra of its metal complexes. The solution spectra of the complexes show an intense transition (log ε= *4.30-4.90)* at 333.3-357.1 nm due to n→π* transition of the carbonyl portion. Although Cl→M(II) charge transfer bands might also be expected in this region of the spectrum [60], it does not add significantly to the magnitude of ε in the various chloro complexes. In the complexes, the d→ d bands are at different energies in solution compared to the solid state. Often two bands in the solid state spectra become a single band in solution due to axial bonding of the solvent. The shifts in the d-d bands to higher energy in DMF compared to the solid state spectra for halo complexes are probably due to partial substitution for one of the halo ligands. The four coordinate nickel (II) complexes have higher energy d-d bands than the corresponding copper (II) complexes indicating stronger metal-ligand bonding, consistent with the metal-ligand bond distances found for analogous nickel(II) and copper(II) complexes with thiosemicarbazide and thiosemicarbazone ligands [61]. Further, the bromo complex $[Cu(H_2L)_2Br_2]H_2O$ has a higher d→d maximum than the corresponding chloro complex $[Cu(H_2L)_2Cl_2]H_2O$ suggesting that the longer Br-Cu(II) bonds allow stronger bonding of the thiosemicarbazide ligand. The solid complex $Fe(H_2L)(HL)Cl_2$ showed two bands at 487.8 nm and 648.9 nm, which are assignable to $^6A_{1g} \rightarrow {}^4A_{1g}(G)$ and $^6A_{1g} \rightarrow {}^4T_{2g}(G)$ transitions, respectively, suggesting its octahedral nature [62]. The magnetic moment of the iron complex is 5.7 BM which is closer to the spin only value indicating an octahedral structure for the complex [63].

The electronic spectra of cobalt(II) complexes $[Co(H_2L)_2Cl_2]H_2O$ and $[Co(H_2L)_2(NO_3)_2].3H_2O$ have spectral typical of an octahedral system. The solid state electronic spectra of the complexes are very similar to each other and show two bands, in the ranges 625.0-649.4 nm and 486.1-469.5 nm, assignable to $^4T_{1g}$ (F) → $^4A_{2g}$ (F) (ν_2) and $^4T_{1g}$ (F)→$^4T_{2g}$ (P) (ν_3) transition [64], respectively, indicating an octahedral structure [65]. The ligand field parameters, Dq and B are, in principle, calculated using first order perturbation theory [66] and the transition energies are given by the equation of lever [67]. In table (4), the band maxima with their assignment and the calculated ligand field parameters are listed. The value of Dq has been calculated from the transition energy diagram using the relation ν_3/ν_2 ratio. The ν_3/ν_2 ratio for the cobalt(Il) complexes lies in the 2.16-2.17 range (Table 4) which is the usual range, for octahedral cobalt(II) complex. The neophaleuxautic parameters, β was obtained using the relation β = B (complex)/B(free ion), where B(free ion) is 971 cm^{-1}. The values of β for these complexes indicate the covalent character of the metal-ligand bond is low. Also, the magnetic moments values (μ_{eff}) of the Co(II) complexes are in the range 5.1-5.3 B. M., which lie in the range reported for high spin (S = 3/2) octahedral cobalt(II) complexes [68]. The electronic spectra of the $[Co(HL)(OOCCH_3)]H_2O$, $[Co(H_2L)Br_2]H_2O$ and $[Co(H_2L)(HL)]ClO_4$ complexes show one main band in the range 665.8-668.0 at attributable to $^2A_{1g} \leftarrow {}^2E_g$ transition in a square planar field [69, 70]. There is also a shoulder at about 500 nm as observed in the case of low spin square-planar complexes. These cobalt complexes displayed a magnetic moment in the range 2.0-2.2 B.M. which correspond to the low spin square planar geometry of the complexes [71].

Table 4. The electronic absorption spectral bands (nm) ligand field parameters for Co(II) complexes

Complexes	ν_2 $^4T_{1g}(F)\rightarrow$ $^4A_{2g}$	ν_3 $^4T_{1g}(F)\rightarrow$ $^4T_{1g}(P)$	ν_1 $^4T_{1g}(F)\rightarrow$ $^4T_{2g}(F)$	Dq	B	β	ν_2/ν_1
$[Co(H_2L)_2Cl_2]H_2O$	648.9	487.8	1406.5	821	966	0.99	2.16
$[Co(H_2L)_2(NO_3)_2].3H_2O$	625.0	469.5	1353.2	853	1003	1.03	2.17

The nickel(1l) complexes, except the $[Ni(HL)_2]H_2O$ have electronic spectra typical of an octahedral system. The solid state electronic spectra of Ni(II) complexes, except the latter one show two bands lie in the range 649.8-626.2 nm and 397.1-393.9 nm which may be assigned to the transition $^3A_{2g}(F)\rightarrow{}^3T_{1g}(F)$ (ν_2) and $^3A_{2g}(F) \rightarrow {}^3T_{1g}(P)$ (ν_3), respectively in octahedral symmetry [72]. The transition $^3A_{2g}(F)\rightarrow{}^3T_{2g}(F)$ (ν_1) is expected to be at 10000 nm. The calculated values of ν_2/ν_1 in the range 1.5-1.6 is consistent with an octahedral configuration [73]. The calculated ligand field parameters for the complexes are listed in table 5.

The values of Dq have been calculated from the transition energy diagram using the relation ν_3/ν_1 ratio. The neophaleuxautic parameter, β, was readily obtained using the relation β = B(Complex)/B(free ion), where B(free ion) equals 1041 cm^{-1}. the values of β for these complexes indicates that the covalent character of value of metal-ligand σ-bond is low. The magnetic moment of $[Ni(HL)_2]H_2O$ equals to 2.6 B. M. lies in the range corresponding to two unpaired electrons [74] and a four coordinated tetrahedral structure [75] is suggested.

The solid state spectrum of $[Ni(HL)_2]H_2O$ showed two absorption bands one at 630.1 nm and well defined band at 833.3 nm which are considered to be due to $^3T_1(F)\rightarrow{}^3A_2$ (ν_2) and $^3T_1(F) \rightarrow {}^3T_1$ (P) (ν_3) transition, respectively for a tetrahedral nickel(II) [76]. The theoretical calculation of Dq, B and β based on this assignment yield 640 cm^{-1}, 750 cm^{-1} and 0.73 respectively. The value of β for t is complex indicates a low covalent character of metal-ligand bond.

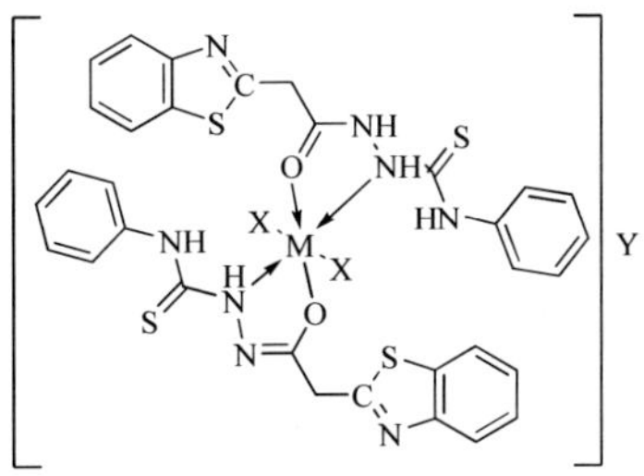

$Fe(H_2L)(HL)Cl_2$, M=Fe; X=Cl; Y= none
$[Co(H_2L)(HL)]ClO_4$, M=Co; X=none; Y= ClO_4

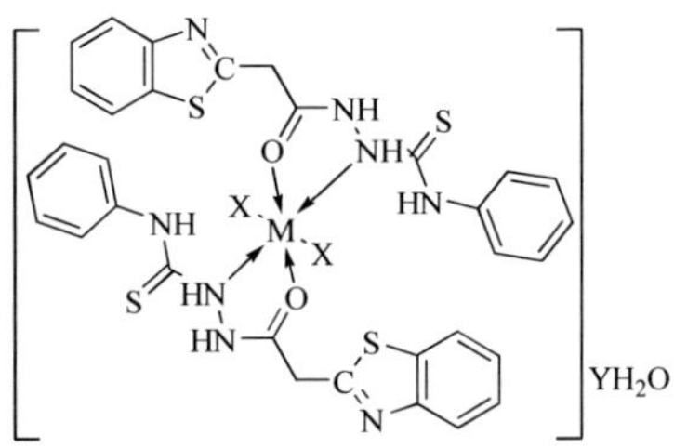

$[Co(H_2L)_2Cl_2]H_2O$, M=Co; X=Cl; Y= 1
$[Co(H_2L)_2(NO_3)_2].3H_2O$, M=Co; X=$NO_3$; Y= 3
$Ni(H_2L)_2(OC(O)CH_3)_2$, M=Ni; X=$OC(O)CH_3$; Y=0
$[Ni(H_2L)_2(NO_3)_2].3H_2O$, M=Ni; X=$NO_3$; Y=3
$[Cu(H_2L)_2Cl_2]H_2O$, M=Cu; X=Cl; Y=1
$[Cu(H_2L)_2(NO_3)_2].2H_2O$, M=Cu; X=$NO_3$; Y=2

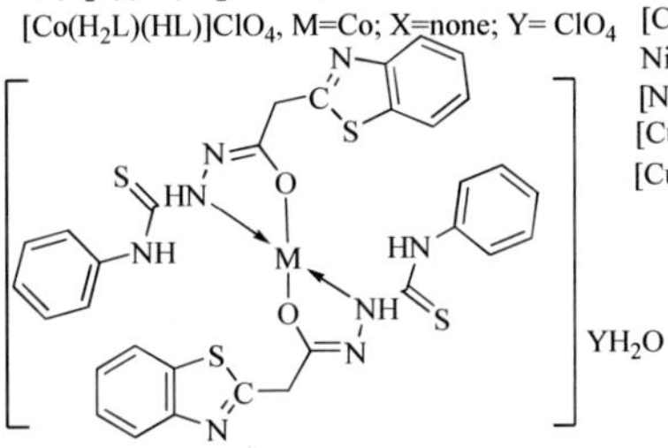

$[Cu(HL)_2(H_2O)_2]H_2O$, M=Cu; Y= 1
$[Zn(HL)_2].3H_2O$, M=Zn; Y= 3

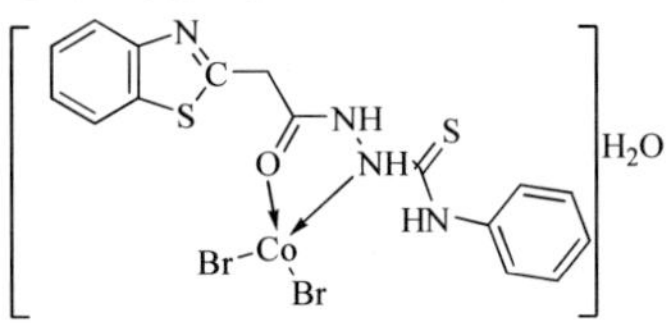

$[Co(H_2L)\ Br_2]H_2O$

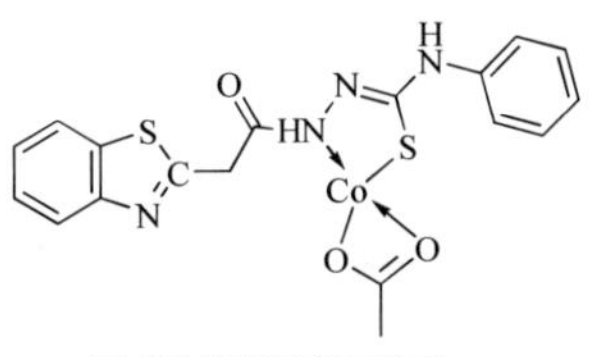

$[Co(HL)(OOCCH_3)]H_2O$

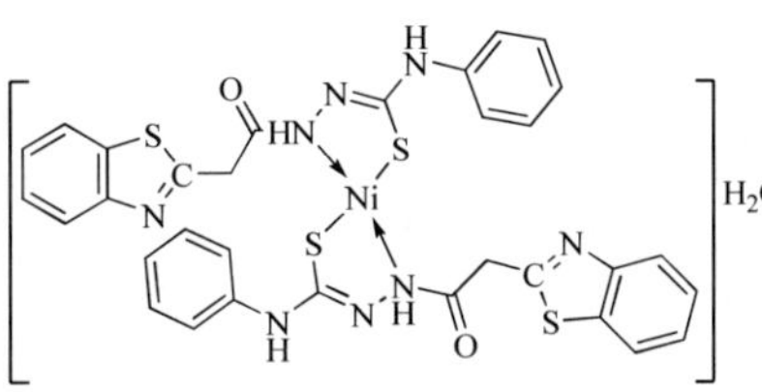

$[Ni(HL)_2]H_2O$

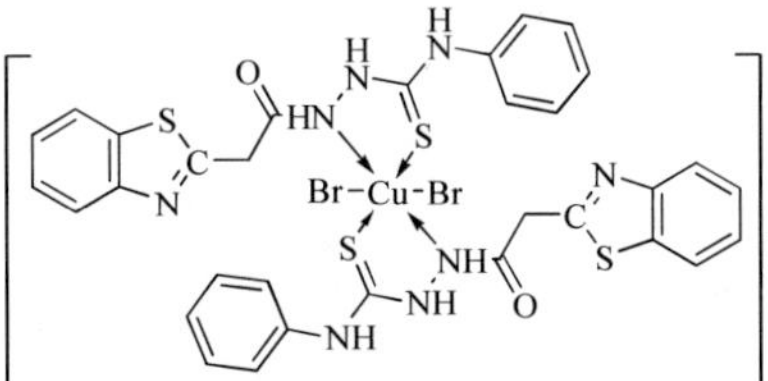

$[Cu(H_2L)_2Br_2]H_2O$

The electronic spectra in nujull mull for all copper(II) complexes are typical and display a broadband in the range 609.8-590.0 nm assigned to superimposed transition $^2B_{1g} \rightarrow {}^2B_{2g} + {}^2E_g$ suggesting a distorted octahedral configuration [77].

Table 5. The electronic absorption spectral bands (nm) ligand field parameters for Ni(II) complexes

Complexes	ν_1 $^3A_{2g}\rightarrow{}^3T_{2g}(F)$	ν_2 $^3A_{2g}\rightarrow{}^3T_{1g}(F)$	ν_3 $^3A_{2g}\rightarrow{}^3T_{1g}(P)$	Dq	B	β	ν_2/ν_1
$Ni(H_2L)_2(OC(O)CH_3)_2$	1048.2	649.8	397.1	954	795	0.76	1.61
$[Ni(H_2L)_2(NO_3)_2].3H_2O$	984.3	626.2	393.9	1015	725	0.70	1.57

REFERENCES

[1] Campbell, J. M., *Coord. Chem. Rev.*, 1975, 15, 279.

[2] Padhye, S. B., Kaufman, G. B., *Coord. Chem. Rev.*, 1985, 63, 1270.

[3] Casas, J. S, Garcia-Tasende, M. S., *Coord. Chem. Rev.*, 2000, 209, 1970.

[4] El-Medani, M. Samir, Omyana, A. M., R.N. Ramaden, *J. Mol. Struct.* 2005, 738, 171.

[5] Zhan. J., Zhan. B., Liu. J., Xu. W.J., Wang. Z., *Spectrochim. Acta, Part A*, 2001, 57, 149.

[6] Liu. W., Zou. Y., Li. Y., Yao. Y., Meng, Q. J., *Polyhedron*, 2004, 23, 849.

[7] Djbbar, S. S., Benali, B. O., Deloume , J. P., *Polyhedron*, 1997, 16, 2175.

[8] Bhattacharyya, P., Parr, J., Ross, A. T., *J. Chem. Soc., Dalton Trans.*, 1998, 3149.

[9] Wu, J.C., Tang, N., Liu, W.S., Tan, M.Y., Chan, A.C.S., *Chin. Chem. Lett.*, 2001, 12, 757.

[10] Pereira, E., Gomes, L. R., Low, J. N., de Castro, B., *Polyhedron,* 2008, 27, 335.

[11] Koley, M. K., Sivasubramanian, S. C., Varghese, B., Manoharan, P. T., Koley, A. P., *Inorg. Chim. Acta*, 2008, 361, 1485.

[12] Ali, M. A., Abu Bakar, H. J., Mirza, A. H., Smith, S. J., Gahan, L. R., Bernhardt, P. V., *Polyhedron,* 2008, 27, 71().

[13] Wang, M., Wang, L.F., Li, Y. Z., Li, Q. X., Xu, Z. D., Qu, D. Q., *Trans. Met. Chem.*, 2001, 26, 307.

[14] Gulerman, N.N., Rollas, S., Erdeniz, H., Kiraj, M., *J. Pharm. Sci.*, 2001, 26, 1.

[15] Tarasconi, P., Capacchi, S., Pelosi, G., Corina, M., Albertini, Bonati, R., A., *Bioorg. Med. Chem.*, 2000, 8, 154.

[16] Charo, J., Lindencrona, J.A., Carlson, L.M., Hinkula, J., Kiessling, R., *J. Virol.*, 2004, 78, 1132.

[17] Mishra. V., Pandeya, S. N., Anathan, S., *Acta Pharm. Turc.*, 2000, 42, 139.
[18] Sakamoto, M., Maneski, K., Okawa, H., *Coord. Chem. Rev.*, 2001, 379, 219.
[19] Kanthimathi, M., Dhathathreyan, A., Nair, B.U., *Chem. Phys. Lett.*, 2000, 324, 43.
[20] Wen, J., Zhao, J., Wang, X., Dong, J., You, T., *J. Mol. Catal. A: Chem.*, 2006, 245, 242.
[21] Bahur, G.Z., Schieitzer, E.Z., *Anorg. Allg. Chem.*, 1955, 278, 136.
[22] Bindu, P., Kurup, M.R.P., Sathyakeerthy, T.R., *Polyhedron,* 1999, 18, 321.
[23] Wada, A., Fujibayshi, Y., Yokoyamma, Arch, A.,. *Biochem. Biophys.*, 1994, 310.
[24] Mathias, C.J., Weich, M.J., Raichie, M.E., Lich, L.L., McGuire, A.H., Zinn, K.R., John, E., Green, M.A., *J. Nucl. Med.*, 1990, 31, 351.
[25] Dearling, J.L.J., Lewis, J.S., Mullen, G.E.D., Welch, M.J., Blower, P.J., *J. Biol. Inorg. Chem.*, 2002, 7, 249.
[26] Christopher, A.W., Yap, P.A.G., Raju, N.P., Greedan, J.E., Crutchley, R.J., *Inorg. Chem.*, 1999, 38, 2548.
[27] Vogel, A.I., *A Text Book of Quantitative Inorganic Analysis, fourth ed., Longman*, London, 1978.
[28] Youssef, N.S., *Synth. React. Inorg. Met.-Org. Chem.*, 2000, 30, 225.
[29] West, D.X., Stark, A.M., Bain, G.A., Liberate, A.E., *Transition Met. Chem.*, 1996, 21, 289.
[30] West, D.X., Salberg, M.M., Bain, G.A., Liberate, A.E., *Transition Met. Chem.*, 1997, 22, 180.
[31] Yulan, Z., Hui, W., Pengfei, X., Shaozu, W., *Synth. React. Inorg. Met.-Org. Chem.*, 1996, 26, 263.
[32] Singh, K., Barwa, M. S., Tyagi, P., *European Journal of Medicinal Chemistry*, 2006, 41(1), 147.
[33] John, R.P., Sreekanth, A., Kurup, M.R.P., Mobin, S.M., *Polyhedron*, 2002, 21, 2515.
[34] Bullock, J.I., Tajmir, H.A., *J. Chem. Soc. Dalton Trans.*, 1978, 36.
[35] Sathyanaray, D.N., Nicholls, D., *Spectrochim. Acta*, 1978, 34A, 263.
[36] Rao, C.N.R., Venkataghavan, R., *Spectrochim. Acta*, 1962, 18, 541.
[37] Colthup, N. B., Daly, L., Wiberley, S. E., *(Introduction to Infrared and Roman Spectroscopy*), Academic Press, New York, 309-327 (1975).
[38] Soliman, A. A., Ali, S. A., Orabi, A., *Spectrochim. Acta Part A*, 2006, 65, 841.

[39] Akinchan, N. T., West, D. X., Yang, Y. H., Salberg, M. M., Klein, T. L., *Transition Met. Chem.*, 1995, 70, 481.
[40] Teotia, M. Gurthu, N., Rama, V. B., *J. Inorg. N ucl. Chem.*, 1980, 42, 821.
[41] EI-Dissouky, A., Fahrny, A., Arner, A., *Inorg. Chim. Acta*, 1987, 133, 311.
[42] Tatwawadi, S. V., Singh, A. P., Narang, K. K., *J. Sci. Res. Banaras Hindu Univ.*, 1980, 3, 143.
[43] Aggarwal, R. C., Singh, N. K., Prasad, L., *Indian J. Chern.*, 1976, 11, 325.
[44] Soliman, A. A., Ali, S. A., Orabi, A., *Spectrochim. Acta Part A*, 2006, *65*(3-4), 841.
[45] West, D. X., Padhye, S. B., Sonawane, S. B., *Structure and bonding*, 1991, 76, 1.
[46] West, D. X., Gebrernedhin, H., Romack, T. J., Liberta, A. E., *Transition Met. Chern.*, 1994, 12, 426.
[47] Nag, S. K., Joarder, D. S., *Can. J. Chern.*, 1976, 54, 2827.
[48] Nakamoto, K., *"Infrared Spectra of Inorganic and coordination compounds"*, Wiley Interscience, New York, P 173 (1965).
[49] Boghaei, D.M., Gharagozlou, M., *Spectrochim. Acta*, 2007, 67A, 944.
[50] El-Shazly, R.M., Al-Hazmi, G.A.A., Ghazy, S.E., El-Shahawi, M.S., El-Asmy, A.A., *Spectrochim. Acta,* 2005, A61, 243.
[51] Nakamoto, K., "Infrared and Raman Spectra of Inorganic and coordination compounds", Wiley Interscience, New York, 1986.
[52] Ferrari, M. B., Fara, G. G., Lafranchi. M., Pelizzi, C., Tarasconi, M., *Inorg. Chern. Acta*, 1991, 181, 253.
[53] Shakir, M., Azim, Y., Chishti, H.-T.-N., Parveen, S., *Spectrochim. Acta Part A,* 2006, *65*(2), 490.
[54] Rahaman, S. H., Ghosh, R., Lu, T.-H., Ghosh, B. K., *Polyhedron,* 2005, 24(12), 1525.
[55] El-Bindary, A.A., *Transction Met. Chem.*, 1997, 22, 381.
[56] Youhong, H., Zienglu, L., Yulan, Z., Shozu, W., *Synth. React. lnorg. Met. Org. Chem.*, 1995, 25, 349.
[57] Soto, L., Sancho, A., Borras, J., *Synth. React. Inorg. Met.-Org. Chem.*, 1986, 16, 595.
[58] Okawa, H., Nakimoto, M., Izumitani, T., Kida, S., *Bull. Chem. Soc. Japan*, 1982, 55, 2671.
[59] Suzuki, M., Kanatomi, I., Demura, Y., Murasc, I., *Bull. Chcm. Soc. Japan*, 1984, 57, 1003.

[60] Mikuriga, M., Okawa, H., Kida, S., *Bull. Chim. Soc. Japan*, 1980, 53, 3717.

[61] Lever, A. B. P., Lewis, J., Nyholm, R. S., *J. Chem. Soc.*, 1963, 2552.

[62] West, D. X., Gebrernedhin, H., Butcher, R. J., Jasinski, J. P., Liberta, A. E., *Polyhedron*, 1993, 11, 2489.

[63] Girish Kumar, K., Saji John, K., *Reactive and Functional Polymers,* 2006, *66*(12), 1427.

[64] Cotton, F.A., Wilkinson, G., *Advanced Inorganic Chemistry*, John Wiley and Sons, New York, 1998.

[65] Tascioglu, S., Yalcin, B., Nasrullayeva, T. M., Andac, O., Buyukgungor, O., Aydin, A., Medjidov, A. A., *Polyhedron*, 2006, *25*(6), 1279.

[66] Lever, A. B. P., "*Inorganic Electronic Spectroscopy*", Elsevier, Amsterdam, 341, (1968).

[67] Tanab, Y., Sugano, S., *J. Phys. Soc. Japan*, 1954, 9, 753.

[68] Lever, A. B. P., *J. Chem. Educ.*, 1968, 45,711.

[69] Masoud, M. S., Haggag, S. S., Ramadan, A. A., Mahmoud, S. A., *Transition Met. Chem.*, 1998, 23, 243.

[70] Ketcham, K.A., Swearingen, J.K., Castineiras, A., Garcia, I., Bermejo, E., West, D.X., *Polyhedron*, 2001, 20, 3265.

[71] Kumar, K. G., John, K. S., *Reactive and Functional Polymers*, 2006, 66, 1427.

[72] Kumar, K. G., John, K. S., *React. Funct. Polym.*, 2006, 66, 1427.

[73] Shakir, M., Azim, Y., Chishti, H.-T.-N., Parveen, S., *Spectrochim. Acta Part A*, 2006, *65*(2), 490.

[74] Hathawy, B. J., Billing, D. E., *Coord. Chem. Rev.*, 1970, 5, 143.

[75] Earnshaw, A., "*Introduction to Magnetic Chemistry*", Acadmic Press, New York, 1968.

[76] Cotton, F. A., Wilkinson, G., "*Advanced Inorganic Chemistry*", Wiley Interscience, New York, P. 761 (1980).

[77] Yumer, M., Koksal, H., Serin, S., *Synth. React. Inorg. Met.-Org. Chem.*, 1989, 26,

In: Smart Nanoobjects
Editor: Kirill Levine

ISBN: 978-1-62808-741-3

Chapter 10

MULTICOMPONENT ELECTROACTIVE AND PH-SENSITIVE SMART COMPOSITES BASED ON POLYPYRROLE, POLYACRYLIC ACID HYDROGELS, AND POLYETHYLENE POROUS FILMS

***G. K. Elyashevich*[*,1], *M. A. Smirnov*[1], *N. V. Bobrova*[1], *I. Yu. Dmitriev*[1] *and V. Bukošek*[2]**

[1]Institute of Macromolecular Compounds,
Russian Academy of Sciences, Saint-Petersburg, Russia
[2]University of Ljubljana, Faculty for Natural Sciences and Engineering
Ljubljana, Slovenia

ABSTRACT

Two- and three-component electroactive and pH-sensitive isotropic bulk composites were prepared by polypyrrole (PPy) synthesis in the crosslinked polyacrylic acid (PAA) hydrogel volume and on porous films. The structure, pH-sensitivity, electric and mechanical properties of the composites were studied. Two different initiators were used to

[*] Corresponding author: elya@hq.macro.ru, Tel: +7 (812) 3286876, Fax: +7 (812) 3286876, Institute of Macromolecular Compounds, Russian Academy of Sciences, 31 Bolshoy pr., 199004 Saint-Petersburg, Russia.

synthesize the PAA gels. The influence of the initiator type on swelling degrees in water media with various acidities were investigated for the PAA hydrogels. Film composites were formed by the PAA hydrogel polymerization on oriented polyethylene (PE) porous films produced in the process based on melt extrusion. The PAA/PE composites were found to be characterized by a higher strength and elasticity than monolityc gels, the swelling degree being the same. The structure and morphological features of the PAA/PPy composites were investigated by SEM and AFM. It was revealed that the composites' structure essentially differed from those of the components: the PAA gel had the structure characteristic of glassy polymers; PPy which was synthesized in the gel was formed in the globular morphology. At the same time, the PAA/PPy composites had a folded surface relief, not typical of pure PAA and PPy. The swelling degree of the PAA/PPy composites was found to increase with growing pH, like for the hydrogel matrix, but the swelling degrees of the composites were higher than that of the matrix. Electroconductivity of the PAA/PPy composites increased with growing PPy content and its value (10^{-3} – 10^{-4} S/cm) was comparable to those for other PPy composites. The viscoelastic properties of the composites studied by DMA in shear deformation proved to be determined by the hydrogel matrix structure, and the PPy phase markedly affected the molecular mobility in the samples. Three-component composites were prepared by forming the PPy phase directly in the reinforced film PAA/PE samples. The PPy/PAA/PE composites had the same swelling degree as two-component PAA/PE, and it was determined by the gel structure. Electroconductivity of the samples along the film surface was 5×10^{-5} S/cm. The elasticity of the PPy/PAA/PE composites proved to be considerably higher than that for the PAA gel and PAA/PPy composites. The electroconducting PAA/PPy composites demonstrated a reversible electromechanical response – bending deformation under the influence of changes in the electric field value or direction. The mechanism of this effect is discussed.

Keywords: Electroactive composites, hydrogels, polypyrrole, crosslinked polyacrylic acid, polyethylene porous films, electroconductivity, pH-sensitivity

1. INTRODUCTION

Conducting polymers and swelling hydrogels are promising materials which can be used as smart systems, such as controlled "soft" sensors, actuators, artificial muscles, and separation membranes. High swelling hydrogels of crossliked polymers are able of not only absorbing water in much larger amounts than their mass but also retaining it under a certain pressure. Their volume considerably changes with varying temperature, pH, and chemical composition of media, and also under electric voltage. From the practical point of view the electric field is the most efficient method of activation because it allows one to control the intensity of external stimuli and to compactly integrate the material into electromechanical systems. However, a low inherent conductivity of a hydrogel polymer network strongly retards its bulk response. The efficiency of electric stimuli can be increased in many times by combining the hydrogel and electroconducting component, i.e., conductive polymer, in one composite [1, 2].

At the same time, a combination of hydrogels and electroconducting polymers in one system at a microlevel is hindered by their insolubility in water and thermodynamic incompatibility of the components. Composite hydrogel/electroconducting polymer systems described in literature [3, 4] are swelling powders sensitive to changes in temperature and chemical composition of the medium or application of electric potential. However, the impossibility to obtain bulk or film samples which are more promising for practical use is a considerable drawback of the methods suggested in literature. Difficulties are caused by a frequently occurring “embrittling” influence of electroconducting polymers on the composite materials into which they are incorporated. Poor mechanical properties are one of the main disadvantages of the conjugated polymers. It imposes considerable limitations on their use in engineering. The problem of preparation of these materials can be solved by using matrices or substrates that can confer mechanical integrity and strength to composites.

The goal of our investigations was to develop methods for preparation of electroactive polymer composites immediately in the bulk state and also in the form of film systems and investigate their mechanical properties, ability to swell in water media with different pH, and also the behavior of these functional systems under electric stimuli. A specific feature of the methods we suggest is that the electroactive component phase is formed in neutral and charged spatially homogeneous media having different natures and structures. These media play a role of nanocontainers for the active component and determine the character of distribution and degree of interconnection of structural elements in two- and three-component systems. The polyacrylic acid hydrogels and polyethylene porous films are used as pH sensitive isotropic and neutral oriented composites with polypyrrole as an electroactive component.

2. EXPERIMENTAL

Hydrogels of crosslinked polyacrylic acid (PAA) were obtained by free radical copolymerization of acrylic acid (AA) monomers (Aldrich). Initiators were two different systems: 1) ammonium persulfate (APS)/ tetramethylethylenediamine – (PAA-1) gels and 2) APS/sodium sulphite/ Mohr's salt - (PAA-2) gels. In both cases the ratio between monomer and initiator was 1000:1.

As a crosslinker, N,N'-methylene-bis-acrylamide (MBAA) in a mole ratio to monomer (AA) of 1:300 was used. The yield of the end product was not less than 98% in all experiments. The value of swelling (*Q*) of PAA hydrogels and composites was determined gravimetrically and calculated as

$$Q = \frac{m_{sw} - m_{dry}}{m_{dry}},$$

where m_{dry} is the sample mass in the dried state, and m_{swell} is the sample mass in the equilibrium swollen state. The experimental error was 5%.

Surface structure and morphological features of the composites and their components were examined by electron microscopy (*JSM*-35 and *JSM*-6060*LV*, *JEOL*, Japan). The hydrogels were freeze-dried to preserve its structure in the swollen state.

Mechanical properties were studied in the static regime of uniaxial extension and in the dynamic regime at shear deformation (*DMA Q*800, *TA Instruments, Delaware, USA*). Square samples 10x10 mm in size with a thickness of 2-3 mm swelled to the equilibrium state were prepared for testing. The experiment was carried out in the static regime and included simultaneous registration of force and displacement at a velocity of 1 mm/min and temperature 25.0 ± 0.5°C.

The measurements were presented in the form of dependences of stress σ (MPa) on relative deformation ε (%). The modulus of elasticity *E* (MPa) was determined from the slopes of the initial parts of deformation curves which were linear up to ε = 5 – 8%. The deformation response of the PAA/PPy composites in electric field was measured in a three-electrode cell in which the sample was a working electrode, a platinum plate was a counter electrode, and a silver-chloride electrode served as a reference one.

Studies of electric properties were performed at room temperature by using an indium-gallium alloy as a contact material. Electrical conductivity in a dc field was determined by measuring resistances of rod like dried samples 2 mm in diameter and 20 mm in length and calculated as

$$\chi = \frac{l}{SR},$$

where l is the length, S is the cross section area, and R is the electric resistance of the sample.

3. Results and Discussion

If initiator 1 is used to prepare PAA gels, synthesis should be carried out in closed systems in the absence of oxygen which is an inhibitor of the copolymerization process. If oxygen is not removed, the activity of initiation decreases, which reduces the reproducibility of the characteristics of the samples obtained. For this reason, the synthesis is performed under a nitrogen flow.

To provide the gel formation process on substrates or in open vessels and to increase the range of methods by which the samples obtained could be investigated, an “ammonium persulfate - sodium sulphite - Mohr's salt” system was used for the first time in this work as an initiator (initiator 2) – in synthesis of the PAA gel.

Owing to the choice of this initiator, reaction could be accomplished in vessels that were not isolated from oxygen of air, and gel could be formed on substrates or obtained as films.

Equilibrium degrees of swelling were determined in a wide range of pH for the hydrogels synthesized with each of two types of initiators. It was found that Q_p grew with increasing pH for both gel types and reached a maximum at an ultimately accessible pH of 11.5. The curves exhibited a rapid growth, i.e., the highest pH-sensitivity, in the region of pH from 6 to 10; at pH > 8 the growth of the swelling degree became slower and showed a trend to a maximum value.

As one can see from Figure 1, the PAA-2 gels have higher degrees of swelling at all pH than PAA-1. Therefore, the use of initiator 2 allowed us not only to extend the area of experimental research, but also to achieve higher pH sensitivities of hydrogels.

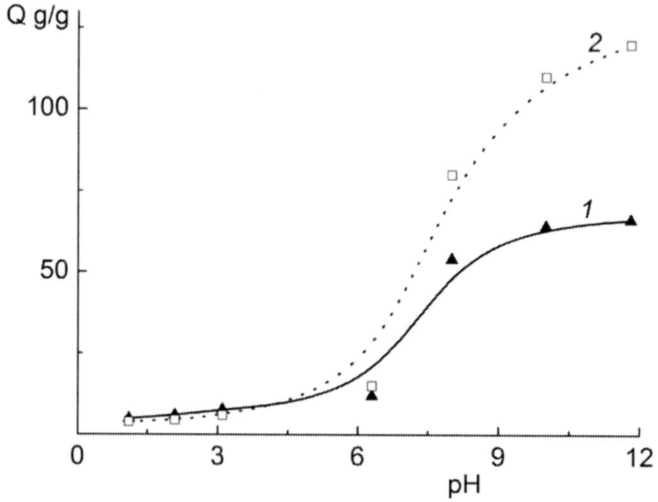

Figure 1. The dependence of hydrogel swelling degree on pH for PAA-1 (1) and PAA-2 (2).

Composite Systems Containing Hydrogel on Porous Polyethylene Films

The drawback of all hydrogel systems is their poor mechanical properties. Because of a low strength in the swollen state (up to destruction at high swelling degrees) and a high brittleness in the dry state, hydrogels are usually studied in the disperse state, i.e., in the form of small or powder-like particles of a millimeter size, while film and bulk materials are more promising for practical applications. Preparation of high swelling hydrogels in the form of a film material is of considerable interest, because the field of their potential applications becomes much more extensive in this case. To solve this problem, in our studies the hydrogels were synthesized directly on a porous polymer substrate, owing to which a system acquired mechanical integrity and sufficiently high strength even at high values of water absorption.

To obtain composite systems with a high swelling hydrogel on a polymer film, the PAA-2 gel was used as an active pH-sensitive component, and the substrate was a porous polyethylene (PE) film formed from commercial HDPE (M_w = 2.5x10^5, M_w/M_n = 10) by melt extrusion with subsequent annealing, uniaxial extension, and thermal fixation [5]. The films were 15 μm thick, the overall porosity was 45%. The porous substrates obtained by this method contained through pores with an average size of 200 nm and had a strongly developed relief-like surfaces (Figure 2), which provided a high adhesion to coatings and active components introduced into them. Pore walls in these films were not continuous-solid, because they were formed by stacks of polymer crystalline elements (lamellae), and so the films contained a three-dimensional system of interconnected pores in which a spatial nano-size active phase was formed [6, 7]. These porous structures based on chemically inert polymers to which polyethylene belongs can serve as ideal "nanocontainers" for incorporating active substances if these substances have important useful properties but are unable of film formation by themselves and cannot be processed by other methods to obtain functional materials.

To prepare samples of PAA-2/PE composites, the PE films were immersed into a polymerization mixture. After hydrogel polymerization, white-milk opaque (due to light scattering from pore walls) PE films became transparent, which indicated that gel was formed not only on the film surface but also in the pore volume. The mass content of gel in the composite was 80-90%, and the composite thickness exceeded in 3-4 times the porous substrate thickness, i.e., the composite was a gel in which a reinforcing porous matrix was incorporated.

Of interest is the fact that the degrees of swelling of the composites proved to be the same as those of a monolithic gel. Of more importance is that the use of a substrate did not adversely affect basic properties of the hydrophilous-hydrophobic reinforced composite.

A significant feature of the porous PE films used as matrices for composite PAA/PE systems is an oriented character of their structure (Figure 2) formed under the conditions of orienting actions at the stages of extrusion and uniaxial extension in the pore formation process, which leads to their different mechanical characteristics in the direction of orientation and in the perpendicular direction. At the same time, hydrogels in the bulk form are mechanically isotropic.

As measurements of mechanical characteristics of the PAA-2/PE composites and their constituents have shown, the PAA-2/PE composites have a much higher strength than that of the gels without substrates (72 and 25 MPa, respectively) and the modulus of elasticity exceeding that of the PE films (3000 and 1000 MPa, respectively). The most important result is that the elasticity of the composites well above that of monolithic gels: the break elongation of the composites reaches 20%, while for dried PAA-2 gels it is as low as 1.5 %. Therefore, owing to a substrate, the PAA/PE composites have a higher mechanical strength and elasticity than a monolithic gel. In addition, in contrast to oriented PE substrates, they exhibit similar properties when deformed in different directions, which is typical of a gel component.

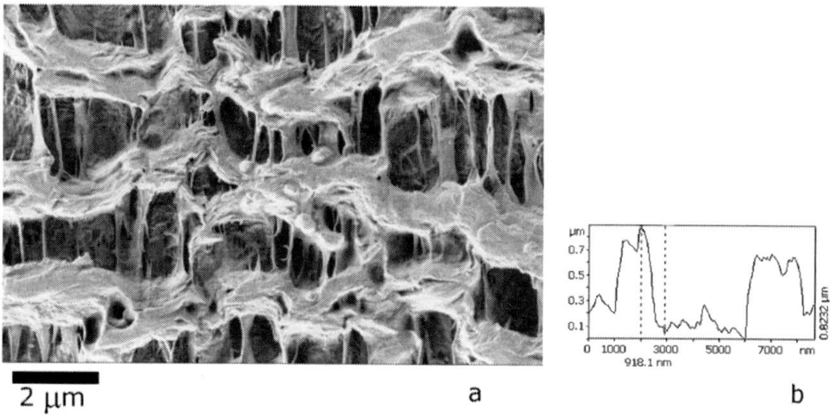

Figure 2. SEM image (a) and relief profile plotted by AFM (b) for porous PE film.

Two-Component Composite PAA/PPy Systems Obtained by Formation of an Electroactive Component in a PAA Hydrogel Matrix

Preparation

To prepare electroactive composite systems, methods were developed and optimal conditions for formation of the conducting component (polypyrrole) phase in the PAA matrix based on *in situ* polymerization of pyrrole directly in the hydrogel bulk [8] were found. Composite PAA/PPy systems were obtained by the two-stage method (Figure 3). The first stage included sorption of pyrrole by a dry PAA hydrogel from a 50-% water/methanol mixture with a pyrrole concentration from 0.05 to 0.3 mol/l during 24 hours (Table 1). The second stage involved oxidation of pyrrole in the gel in a water solution of APS with a concentration of 0.01 mol/l during 1 hour. The sample rapidly transformed from transparent to black, which is the color of polypyrrole. The degree of pyrrole conversion in the gel phase was not lower than 95%.

The weight content of conducting polymer in the dried PAA/PPy samples was calculated as

$$x_{PPy} = \frac{m_{PAA/PPy} - m_{\mathrm{PAA}}}{m_{PAA/PPy}},$$

where $m_{PAA/PPy}$ is the mass of a dried composite PAA/PPy system, and m_{PAA} is the mass of a sample of a dried initial PAA hydrogel.

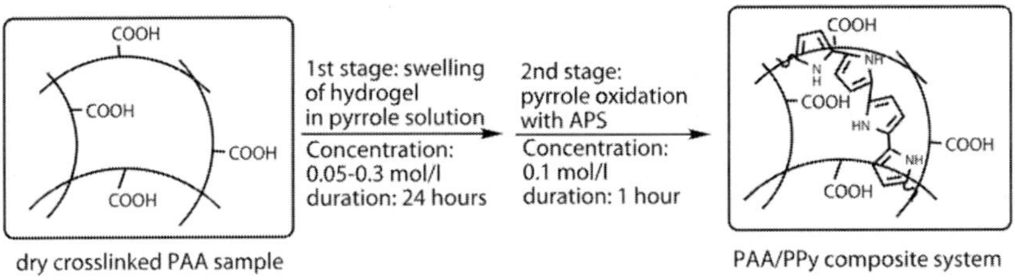

Figure 3. Schematic image of two-stage preparation method of PAA/PPy composites.

Raman scattering spectroscopic data obtained for the composite systems show that polypyrrole formed in the gel phase has a chemical structure similar to that of PPy obtained by a conventional method of oxidative polymerization of pyrrole in a water solution.

Structure and morphological features of the PAA/PPy samples obtained were examined by scanning electron microscopy (SEM). The SEM image of the PAA hydrogel after freeze-drying (Figure 4a) reveals a porous structure of the matrix with a pore size of 3-5 μm. Figure 4b shows that the electroconducting polymer is in the form of globules on the inner surface of hydrogel pores.

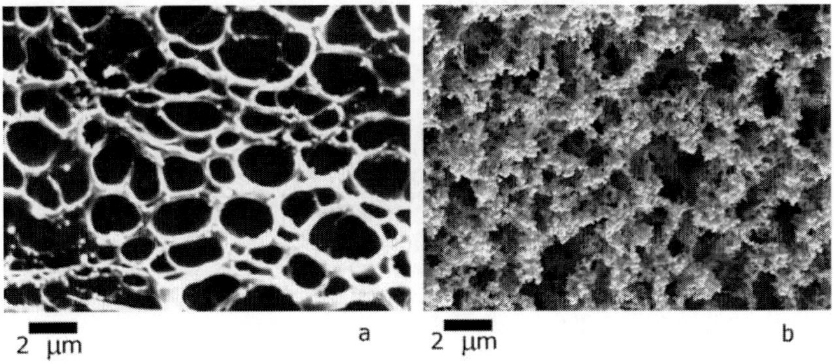

Figure 4. SEM images of initial PAA matrix (a) and PAA/PPy composite (b). Pictures were made from freeze-dried samples.

The SEM images presented in Figure 4b demonstrate the hydrogel volume filling with polypyrrole particles. It can be seen that polypyrrole is uniformly distributed in the PAA matrix space, the globular polypyrrole formations fully masking the structure of the initial hydrogel.

It is also evident from Figure 4 that the composite's structure considerably differs from the structures of initial components (Figure 5 a-c). The PAA surface (Figure 5a) does not contain any specific features and is typical of glassy polymers. PPy synthesized in the absence of matrix by using the same method as for the composite demonstrates a globular morphology (Figure 5b).

At the same time, the morphology of the composite is characterized by a folded surface relief, not typical of pure PAA and PPy (Figure 5c). This indicates that a new morphological structure that includes both components of the system which are structurally combined at a nano-scale level is formed in the composites.

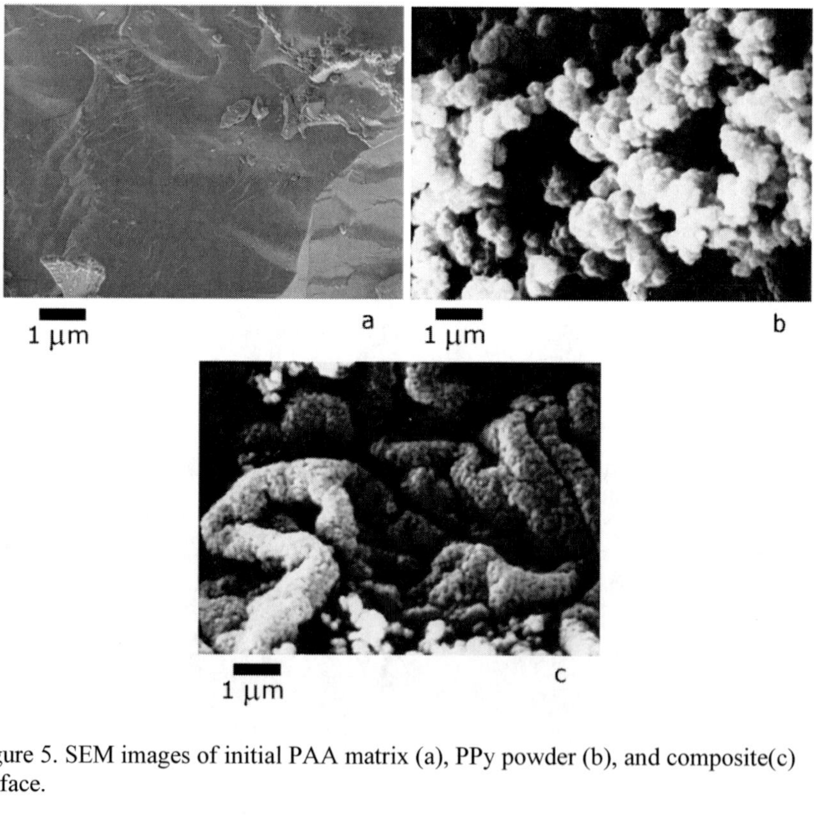

Figure 5. SEM images of initial PAA matrix (a), PPy powder (b), and composite(c) surface.

Swelling in Water Media

The influence of pyrrole concentrations on the degree of swelling of composite systems in water at the first stage of synthesis was studied. The data listed in Table 1 show that an increase in the pyrrole concentration at the first stage leads to a growth in the content of conducting polymer X_{PPy} in the composite system. A decrease in the PPy content at high pyrrole concentrations can be due to a partial destruction of the network structure of the matrix resulting from formation of a rigid framework of polypyrrole in the composite. The effect of rupture of crosslinks under the conditions of PPy synthesis is confirmed by the measurements of modulus of elasticity of PAA hydrogels and PAA/PPy composites (see Section "*Mechanical properties*").

The degrees of swelling in water were measured for all the composites obtained (Table 1). It was found that, if the polypyrrole content is low, composite systems have approximately the same degrees of swelling as the PAA matrix, regardless of the conditions of polypyrrole synthesis.

An increase in the PPy content in the composite causes an expected decrease in the degree of swelling, which is due to formation of a rigid framework of the electroconducting phase that hinders changes in the system sizes, thereby limiting matrix swelling. However, at pyrrole concentrations higher than 0.2 mol/l, the degree of swelling somewhat increases because of chain breaking in the hydrogel network.

Table 1. Properties of composite PAA/PPy systems synthesized at different pyrrole concentrations

$C_{pyrrole}$, Mol/l	X_{PPy}, mass. %	Q, g/g	χ, S/cm
0.10	12	60	10^{-5}
0.15	15	50	2×10^{-4}
0.20	18	36	8×10^{-4}
0.25	17	38	3×10^{-3}
0.30	16	39	5×10^{-3}

The effect exerted by the pH of the medium on swelling of PAA/PPy composites was also investigated. Depending on pH, the degree of swelling Q of the composites sharply increases in alkaline media (Figure 6), the dependences being similar to those for the initial hydrogel matrices (Figure 1). However, the absolute value of Q for the composites is even somewhat higher than for the PAA matrices in the entire range of pH studied: the degree of swelling in water for matrices was 20, and for composites it was 25.

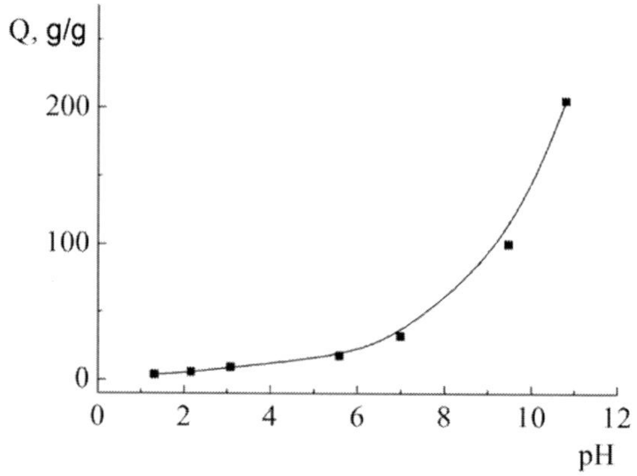

Figure 6. Swelling degree of composites PAA-1/PPy as a function of pH.

Electric Properties

Electrical conductivity (χ) in the DC regime was measured for the PAA/PPy composites as a function of pyrrole concentration in synthesis. Table 1 lists electrical conductivities of the samples synthesized at an oxidizer concentration of 0.01mol/l. As pyrrole concentration increases, the electrical conductivity grows, which is understandable because this growth accompanies the overall increase in the conducting phase content in the composite system. At a pyrrole concentration of more than 0.2 mol/l, the growth of electrical conductivity slows down, which is due to saturation of the composites with PPy that occurs in this range of pyrrole concentrations. Note that the growth in the oxidizer concentration (APS) from 0.01 to 0.02 mol/l leads to a decrease in the conductivity of composites despite the increase in the PPy content. This is the consequence of the destructive oxidizer effect on the PPy structure that reduces the conjugation length.

The electrical conductivities of 10^{-3} – 10^{-4} S/cm achieved in our investigations are comparable to those reported in the literature for the composites of this type [7-9]. It should be noted that the composites with the highest level of electrical conductivity are also characterized by the degrees of swelling in water amounting to about 40, which is rather high for such systems.

The method of electroconducting phase formation in the hydrogel bulk we developed allowed us to prepare bulk composite systems combining the electrical conductivity of polypyrrole and the PAA ability of many-fold changes in sizes at swelling in water and water electrolytes. The investigations have shown that changes in the PAA matrix associated with partial destruction of its crosslinked structure occur during synthesis of PPy. The character of PPy distribution and its volume content depend on pyrrole concentration in synthesis and exert a decisive influence on the composite swelling and its electrical conductivity.

Mechanical Properties

Viscoelastic properties of the composites were examined by dynamic mechanical analysis in the regime of shear deformation with a constant oscillation amplitude of 2 μm in the frequency interval 1-100 Hz in the isothermal regime at 25°C. The behavior of the DMA curves for the PAA and PAA/PPy samples (Figure 7) indicates that the mechanical response in these systems is determined first of all by viscoelastic properties of the PAA matrix, while the presence of the polypyrrole phase manifests itself only slightly in the values of the real part of the shear modulus of the composite. However, the PPy phase has a pronounced effect on the molecular mobility in the composites.

For all the samples, a region of rubber elasticity at frequencies lower than 40 Hz and the viscoelastic behavior in the frequency interval 40-80 Hz (at room temperature) are observed. At frequencies above 75-80 Hz, a decrease in the modulus G\` (Figure 7a) occurs. This is caused by fracture of the samples (or their disintegration) resulting from "mechanical glass transition" of the PAA matrix.

As one can see from Figure 7b, the value of the PAA glass transition at 57-62 Hz is less for the composites than for the initial matrix, and it decreases with increasing PPy content. This is likely to be due to a limitation of mobility resulting from formation of the PPy phase. In addition, the decrease in the intensity of this peak for the composites prepared at a pyrrole concentration of 0.05-0.25 mol/l (curves 2-4, Figure 7b) is accompanied by its narrowing, while the width of the peak for the composite with the highest concentration $C_{pyrrole}$=0.30 mol/l (curve 5) is comparable to that for the PAA matrix.

These features point to the existence of at least two competing factors responsible for the observed changes in the mechanical relaxation spectrum of the composites. On the one hand, the PPy phase enhances the rigidity of the composites, on the other hand, the processes of formation and accumulation of PPy in the hydrogel matrix lead to destruction of the existing crosslinked PAA structure.

The supposition that the frequency of knots of the PAA network becomes lower when the PPy phase is formed is confirmed by the observed decrease in the modulus G\` with increasing PPy content in the PAA/PPy composites (curves 2-5, Figure 7a). Thus, the conducting polymer exerts a complicated influence on the elastic properties of the gel by fixing its structure in the swollen state, on the one hand, and causing a partial destruction of the crosslinks, on the other hand.

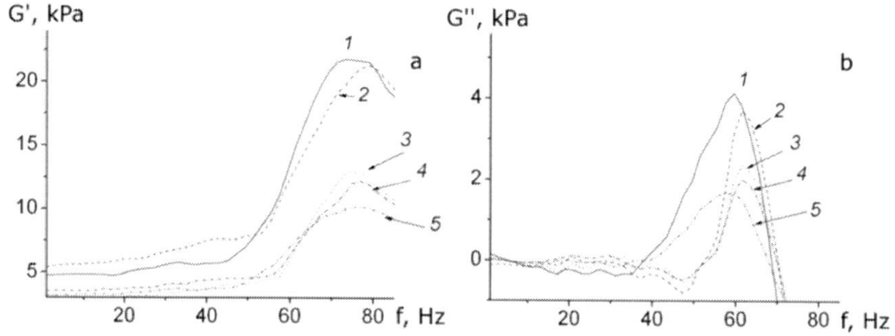

Figure 7. The dependences of real (a) and imaginary (b) parts of shear modulus on frequency for PAA hydrogel (1) and PAA/PPy composites synthesized at concentrations of pyrrole C_{Py}: 0.05 (2), 0.2 (3), 0.25 (4) and 0.3 (5) mol/l.

Three-Component Electroactive Systems Based on Porous PE Films

An important task from the practical point of view is to obtain electroactive and pH-sensitive materials in the film form. We developed methods for preparation of such materials – three-component composite systems – by forming the electroconducting component phase (PPy) immediately in the reinforced film hydrogel – PAA/PE. To prepare such systems, the PAA/PE samples were treated by a pyrrole solution and then by an oxidizer, similar to the method of synthesis of two-component PAA/PPy systems. The film composites obtained had a black color typical of polypyrrole, they were able to swell in water solutions and had an electrical conductivity of 5×10^{-5} S/cm along the film surface. The swelling degree of the electroactive PAA/PPy/PE composites was nearly equal to that of two-component PAA/PPy samples and was determined by the gel structure.

Tables 2 and 3 list mechanical characteristics of the samples of the composites obtained. It is evident that the characteristics of the PAA/PE samples measured at uniaxial extension differ from the characteristics of PAA and PE components (Table 2). At low deformations (to 2%) a high elastic modulus (E) of the composites is provided by the PAA layer, at high deformations the modulus decreases to that of the PE substrate. The substrate also confers to the PAA/PE composites the strength (σ) three times as high as the PAA gel strength. The most pronounced effect is exerted by the substrate on the elasticity of the composites, i.e., they are characterized by the break elongation (ε) nearly equal to that of the substrate, while for the PAA gel this value is more than an order of magnitude lower. Table 3 presents mechanical characteristics of electroactive two- and three-component samples measured in the regime of compression. As one can see from Table 3, the break elongation ε for the PAA/PE/PPy composite appreciably exceeds that of the PAA gel and PAA/PPy composite and amounts to 10%. This is ensured by the elastic PE substrate. Its effect in this case is also stronger than the effect of the rigid-chain PPy that typically reduces elasticity. In contrast to oriented mechanically anisotropic PE substrates, the composites based on them had similar mechanical characteristics in both directions and the break elongations comparable to the maximum break elongations for substrates. They exhibited rather high breaking strength, and their modulus of elasticity was considerably higher than that of the substrates.

Table 2. Mechanical characteristics of porous PE substrates, PAA gel films, and composite PAA/PE films

Sample	σ, MPa	ε, %	*E*, MPa
PE *	110/22	23.0/6.6	880/1200
PAA	25	1.5	5000
PAA/PE	72	20.0	2880

*) numerator – in the orientation direction, denominator – normal to the orientation direction.

Table 3. Properties of electroactive composite PAA/PPy and PAA/PPy/PE systems

Sample	X_{PPy}, mass.%	*Q*, г/г	χ, S/cm	*E*, MPa	ε, %
PAA/PPy	24	11	$10^{-3} - 10^{-4}$	3000	1.5
PAA/PPy/ PE	20	12	5.0×10^{-5}	2900	10

Electric Stimuli Operation

The behavior of PAA/PPy functional systems upon electric stimuli has been investigated. For the experiments, a sample in the film form about 1 mm thick was used. The composite was in the equilibrium swollen state in the medium of 1M water solution of NaCl. The PPy content in the composite was 18%, the electrical conductivity of the composites reached 0.8 S/cm at a degree of swelling in the working solution of 60.

When the electric field magnitude or direction was changed, the composites demonstrated a reversible electromechanical response – bending deformation (Figure 8). The maximum deformation of the sample reached 15 degrees under a potential of 1.5 V and developed during ~2 minutes. In the case of anode connection of the sample, its bending towards the counter electrode was observed (Figure 8a). The anode process at the polymer electrode was accompanied by PPy oxidation, during which Cl^- anions were actively incorporated into the PPy chain to maintain the electroneutrality.

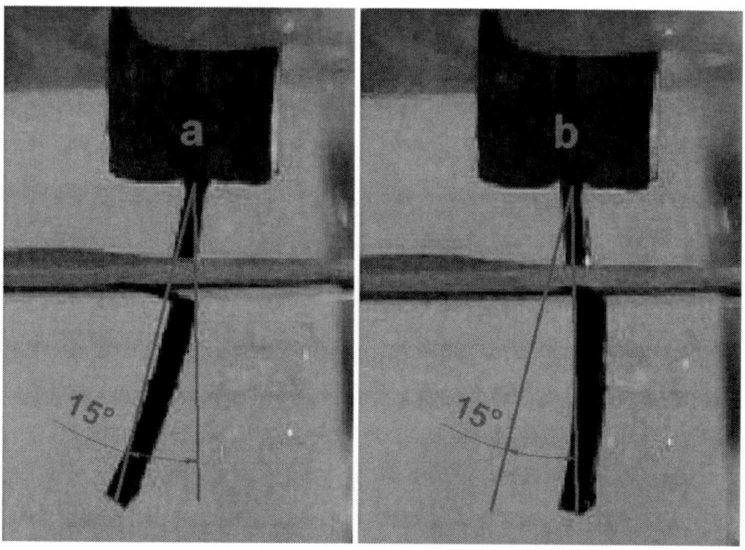

Figure 8. Photographs, demonstrating the bending deformation of the PAA/PPy sample upon electric field at 1V when the sample is an anode (a) and the sample is a cathode (b).

In the case of cathode connection, the sample bending occurred in the direction opposite to the counter electrode (Figure 8b), which was accompanied by the reduction reaction in the PPy phase [12]. It is necessary to note that the bending deformation of the composite could be due to both the processes proceeding in the PPy phase and changes in the hydrogel phase volume. It is known [13] that the deformational behavior of the PAA hydrogel in the electric field depends on the character of transport of mobile ions in the sample volume and in the surrounding solution, which eventually leads to conformational changes in the crosslinked gel structure. The results obtained in our studies have shown that the PAA/PPy composite can be regarded as a promising material for the use as electrical stimuli responsible systems.

CONCLUSION

Thus, methods for preparing two- and three-component bulk and film pH-sensitive and electroconducting systems having a high adhesion between components and preserving a mechanical integrity under deformational actions have been developed. A specific feature of these systems is a spatial continuity of the phase of functionally active component formed in the matrix that contains a system of interconnected pores. It determines the structure of the system, mechanism of formation of the active phase, and its sensitivity to external factors. The investigations have shown that the structure of the active component phase that confers "smart" properties to the multicomponent system determines characteristics of nanocomposites. An important advantage of the methods developed is that composite systems can be prepared in the form of films, which considerably extends the possibilities of their investigations and applications.

ACKNOWLEDGMENTS

The work was supported by the RFBR (grants 10-03-00421-a and 11-03-12054-obr-m).

REFERENCES

[1] Ali, T.O. *Surface and Coatings Technology* 2006.vol. 200, № 12-13 3918- 3925.

[2] Guo, M.Y.;Wei, Y.; MacDiarmid, A.G.; Lelkers, P.I. *Biomaterials.* 2006, vol. 27, № 13,. 2705- 2715.

[3] Han, G.; Shi, G. *Thin Solid Films.* 2007, vol. 515, № 17, 6986- 6991.

[4] Wu, Y.; Alici, G.; Sprinks, G.M.; Wallace G.G. *Synth. Met.* 2006, vol. 156, № 16-17, 1017-1022.

[5] Elyashevich, G.K.; Rosova, E. Yu.; Karpov T.A. Microporous polyethylene film and method for its preparation. Patent of Russian Federation № 2140936, 1997.

[6] Elyashevich, G.K.; Karpov, E.A.; Kozlov, A.G. *Macromol.Symp.* «Mechanical Behavior of Polymeric Materials». Ed.: J.Kahovec, 1999, vol. 147, 91- 101.

[7] Elyashevich, G.K.; Smirnov, M.A.; Kuryndin, I.S.; Bukosek, V. *Polym. Adv. Technol.*, 2006, vol.17, 700-704.

[8] Smirnov, M.A.; Bobrova, N.V.; Dmitriev, I. Yu.; Bukošek, V.; Elyashevich, G.K. *Polym. Sci.*, Ser. A. 2011, vol. 53, № 1, 67-74.

[9] Gangopadhyay, R.; De, A. *Synth. Met.* 2002, vol. 132, № 1. 21-28.

[10] Chakraborty, M.; Mukherjee, D.C.; Mandal, B.M. *Synth. Met.* 1999, vol. 98. № 3, 193-200.

[11] Zoppi, R.A.; Paoli, M.-A.D. *Polymer.* 1996, vol. 37, № 10, 1999- 2009.

[12] Otero, T.F.; Boyano, I.; Cortés, M.T.; Vázquez, G. *Electrochimica Acta.* 2004, vol. 49, 3719-3726.

[13] Shiga, T. Adv. in *Polym.Sci.* 1997, vol.134, 131-163.

In: Smart Nanoobjects
Editor: Kirill Levine
ISBN: 978-1-62808-741-3

Chapter 11

Evaluation of Antifibrotic Activity of Newly Synthesized Pyrimidine Derivatives and Their Metal Complexes

Fathy A. El-Saied[1*], Tarek A. Salem[2], Atef A. Hamed[1], Ahmed S. Radwan[1] and Hussain H. Alganzory[1]

[1]Chemistry Department, College of Science, Qassim University, KSA, Saudi Arabia

[2]Biochemistry Department, College of Medicine, Qassim University, KSA, Saudi Arabia

Abstract

Nickel(II), cobalt(II) and copper(II) complexes with two new pyrimidine monoxime derivatives were prepared and characterized using a variety of analytical and spectral methods. The geometrical shapes for the prepared metal complexes were also elucidated. We tested their possible activities to inhibit liver fibrosis induced by carbon tetrachloride

[*] Corresponding author: Prof. Fathi A. El-Saied, Professor of Chemistry, College of Science, Qassim University, Duriaydah city, Qassim, KSA. Tel.: +966 502446907. E-mail. fathi_elsaied@yahoo.com

(CCl_4) in rat model. Liver function, hepatic hydroxproline (HPO), antioxidant enzymes and serum level of TNF-α and fibronectin (Fbn) were assayed to evaluate the antifibrosis activity of these complexes. Results revealed that administration of different derivatives of complexes to the CCl4-intoxicated rats led to amelioration in the hepatic antioxidant capacity, where, MDA was significantly decreased in parallel with augmentation of catalase, SOD and GSH. In addition, levels of fibrogenic markers, HPO, TNF-α and Fbn were significantly diminished as compared with CCl4- intoxicated rats ($p>0.005$). The antifibrotic activity of pyrimidine complexes was observed to be relied on the types of both ligand and metal ion. Complexes with ligand HL^2 exhibited a potent activity in reducing fibrosis and this may be referred to the presence of phenyl ring as side-chain which has a potent free radical scavenging activity instead of propylene group in the complex1. Also, the antifibrotic activity of pyrimidine monoxime complexes was affected by the type of metal ions.

Keywords: Pyrimidine monoxime, Antifibrosis, Antioxidant, Fibronectin

INTRODUCTION

Liver fibrosis results from chronic damage to the liver in conjunction with the progressive accumulation of fibrillar extracellular matrix proteins. The main causes of liver fibrosis are infection with hepatitis B or C, alcohol abuse, and non-alcohol steatohepatitis. Hepatic fibrosis is characterized by increased deposition and altered composition of extracellular matrix (ECM). Its final stage is cirrhosis, with the liver architecture distorted by collagen bands and formation of islands of regenerating parenchymal cells [1].

Many of the cellular mechanisms have been associated to hepatic fibrosis. Cytokines are soluble autocrine and paracrine mediators. Expression of several cytokines has been described in human liver diseases and experimental liver injuries [2]. Tumor-necrosis factor-alpha (TNF-α) and interleukin-6 (IL-6) are considered major hepatotoxicity mediators in several experimental models of liver injuries [2].

Fibronectins are high molecular weight glycoproteins that exist as soluble dimmers circulating in the blood and other tissue fluids, or as insoluble multimeric fibrils that are incorporated into the extracellular matrix. It was suggested that the accumulating level of cellular fibronectin after long-term ethanol administration in a rat model may be an early response to liver injury, and could lead to the activation of a fibrogenic response in hepatic stellate cells [2].

Carbon tetrachloride (CCl_4) is a hepatotoxin, causing liver necrosis, fibrosis and cirrhosis when administered sequentially. CCl_4-induced fibrosis shares several characteristics with human fibrosis of different etiologies; thus, it is an adequate model of human fibrosis. CCl_4 induced liver damage had been thought to depend on the formation of reactive intermediates such as trichloromethyl ($CCl_3^{\bullet}$) free radical produced by cytochrome P 450 mixed function oxidase system especially, P 450-2E1, the major human enzyme responsible for carbon tetrachloride bioactivation [3] and further converted to a peroxy radical ($CCl_3O_2^{\bullet}$). these free radicals react with polyunsaturated fatty acids to propagate a chain reaction leading to lipid peroxidation or bind covalently to lipids and proteins, there was also evidence of binding to DNA, resulting in destruction of membrane structure [4].

Candesartan cilexetil (CC) is a potent, orally active and selective angiotensin II (Ang II) type one receptor blocker. Ang II is the primary effector molecule of the rennin-angiotensin system (RAS), candesartan (CV-11974) is believed to reduce blood pressure by a minimum of intervention in this endogenous system. CC is a pro-drug that is transformed during absorption to the active compound, candesartan, which is further metabolized to the inactive MII (CV-15959) [5]. Previous studies have suggested that the rennin-angiotensin system (RAS) in addition to having role in regulation of blood pressure and body-fluid homeostasis is involved in hepatic fibrogenesis [6].

However, no effective treatment that delays disease progression and complications has yet been found. Recent trials have been conducted to investigate the antifibrotic activity of many natural and synthetic products [7]. Some metal complexes are known to exhibit remarkable biological activities and the efficacies of some therapeutic agents are known to increase upon co-ordination [8, 9].

Some metal-based antibiotics such as bleomycin, streptonigrin, and bactracin have gained recognition and are more effective than pure drugs [10]. Cobalt complex with histidine ligand showed the significant antibacterial and antifungal activity in comparison with commercial antibiotics [11]. The development of metal-based drugs with promising pharmacological application has been reported as transition metals can interact with a number of negatively charged molecules due to different oxidation states they possess [12, 13].

Therefore, the present study was designed to evaluate the effects of newly synthesized different derivatives of pyrimidine complexes on hepatic fibrosis induced in rats by CCl_4, and to investigate their possible mechanisms.

MATERIALS AND METHODS

Chemicals and Instruments

Reagent grade chemicals were used without further purification. Elemental Micro analyses [C, H, N, S] were determined using Elemental Euro -3000. The IR spectra were measured using KBr discs on FT-IR Shimadzu IR prestige-21 spectrophotometer. Electronic spectra in solid states were recorded in Nujol mulls using Varian UV- 1650 pc uv-vis spectrophotometer. The conductance measurements of complexes in ($1x10^{-3}$M) DMF solution were measured at 25°C with a CON510 OAKTON. Magnetic susceptibilities were measured at 27°C using the modified Gouy method with a Johnson Matthey balance. Diamagnetic correction was made using Pascal's constants.

Analytical grade of carbon tetrachloride (CCl_4) (Sigma, C5335) was freshly dissolved in corn oil (1:4; v:v) and kept at RT for further use. Candesartan cilexetil (CC) was prepared in distilled water [14].

Preparation of Ligands HL^1 and HL^2

The ligands, 2-[N-(2-hydroxyimino-1-methyl-propylidene)-hydrazino] -3- allyl- 5,6,7,8- tetrahydro- 3H- benzo [4,5] thieno [2,3-d] pyrimidine-4-one (HL^1) and 2-[N-(2-hydroxyimino-1-methyl-propylidene)-hydrazino]-3-phenyl-5,6,7,8-tetrahydro-3H-benzo[4,5]thieno[2,3-d] pyrimidine-4-one (HL^2) were prepared by mixing an equimolar amounts of compounds 3-allyl-2-hydrazino-5,6,7,8-tetrahydro-3H-benzo[4,5]thieno[2,3-d]pyrimidin-4-one or 3-phenyl-2-hydrazino-5,6,7,8-tetrahydro-3H-benzo[4,5]thieno[2,3-d]pyrimidin-4-one [15, 16] and diacetyl monoxime dissolved in ethanol. The reaction mixture is refluxed for 5 hours and left to cool.

The precipitated product is filtered off, washed several times with ethanol and dried. The product is recrystallized from ethanol and dried under vacuum in presence of anhydrous $CaCl_2$. The chemical formulation of ligands is shown in figure (1).

Table 1. Elemental analysis, color, molar conductivity and molar ratio for ligands and their metal complexes

No	Compounds	Color	Λ_M	Salt	Molar[a] ratio	Elemental analysis				Cl%
						C %	H %	N %	S %	
1	HL^1	Faint yellow	-	-	-	56.4 (56.8)	5.7 (5.9)	20.2 (19.5)	8.4 (8.9)	-
2	$[Ni(L^1)_2]$	red	10.00	$NiCl_2$ $Ni(OAc)_2$	1:1:1	52.8 (52.7)	5.4 (5.2)	17.6 (18.1)	7.6 (8.3)	Nil
3	$[Ni(HL^1)Cl_2(H_2O)].3H_2O$	Dark brown	15.00	$NiCl_2$	1:1:0	36.7 (36.4)	4.8 (5.2)	11.9 (12.5)	5.4 (5.7)	12.2 (12.7)
4	$[Ni(HL^1)(NO_3)_2(H_2O)].H_2O$	Brown	13.60	$Ni(NO_3)_2$	1:1:0	35.3 (35.3)	4.4 (4.3)	16.3 (17.0)	4.8 (5.5)	-
5	$[Co(L^1)_2]$	Brown	5.20	$Co(NO_3)_2$, $Co(OAc)_2$ $CoCl_2$	1:1:0 1:1:0 1:1:1	53.2 (52.7)	4.8 (5.2)	18.9 (18.1)	7.9 (8.3)	Nil
6	$[Cu(HL^1)Cl_2].H_2O$	Brown	3.50	$CuCl_2$	2:1:0	39.2 (38.5)	4.4 (4.7)	12.8 (13.2)	5.6 (6.0)	12.9 (13.4)
7	$[Cu(L^1)_2]$	Brown	6.70	$Cu(OAc)_2$	1:1:0	51.7 (52.3)	4.9 (5.1)	17.5 (18.0)	7.7 (8.2)	-
8	$[CuL^1(OAc)(H_2O)_2]$	Brown	9.51	$Cu(OAc)_2$	1:1:1	44.7 (44.1)	4.7 (5.2)	12.8 (13.6)	5.9 (6.2)	-
9	$[Cu_2L^1Cl_3(H_2O)]$	Brown	13.07	$CuCl_2$	1:1:0	34.3 (33.5)	3.1 (3.6)	12.2 (11.5)	4.9 (5.3)	16.9 (17.5)
10	$[Cu_2L^1(NO_3)_3(H_2O)]$	Brown	17.51	$Cu(NO_3)_2$	1:1:0	30.2 (29.6)	3.0 (3.2)	15.9 (16.3)	-	-
11	HL^2	Pale yellow	-	-	-	59.9 (60.8)	5.3 (5.3)	18.2 (17.7)	7.8 (8.1)	-
12	$[NiL^2Cl]$	Brown	10.13	$NiCl_2$	1:1:0 1:1:1	50.4 (49.1)	4.6 (4.1)	14.9 (14.3)	5.9 (6.6)	7.5 (7.3)
13	$[NiL^2 (NO_3)]$	Yellow	13.11	$Ni(NO_3)_2$	1:1:0	46.9 (46.6)	4.2 (3.9)	16.3 (16.3)	5.8 (6.2)	-
14	$[NiL^2(OAc)]$	Brown	8.36	$Ni(OAc)_2$	1:1:0	51.4 (51.6)	4.2 (4.5)	14.4 (13.7)	5.9 (6.3)	
15	$[Co(L^2)(NO_3)]$	Brown	7.19	$Co(NO_3)_2$	1:1:1	47.1 (46.6)	3.5 (3.9)	16.7 (16.3)	6.5 (6.2)	
16	$[Co(L^2) Cl]$	Violet- red	5.15	$CoCl_2$	1:1:1	50.0 (49.2)	4.1 (4.1)	14.8 (14.3)	5.9 (6.6)	7.5 (7.3)
17	$[CoL^2 (OAc)]$	Brown	13.14	$Co(OAc)_2$	1:1:0 1:1:1	50.9 (51.6)	4.4 (4.5)	14.3 (13.7)	6.4 (6.3)	
18	$[Cu(L^2)(OH)].H_2O$	Green	7.19	$CuCl_2$	1:1:0	48.0 (48.7)	4.1 (4.7)	15.0 (14.2)	8.9 (6.5)	Nil
19	$[CuL^2(OH)_2]$	Pale Brown	15.13	$Cu(NO_3)_2$	1:1:0	49.3 (48.7)	4.3 (4.7)	14.7 (14.2)	5.9 (6.5)	-

a= metal: ligand: NaOH.

Λ_M = molar conductivity, $ohm^{-1}cm^2mol^{-1}$.

Table 2. Infrared spectral bands (cm^{-1}) and their assignment for ligands and their complexes

No	υ H_2O	υ (OH)	υ (N-H)	υ(C=O)[b]	υ(C=N) imine	υ(C=N) oxime	υ (N- O)	υ(OAc)	υ(NO_3)	υ(M-N)
1	-	3333	3237, 3310	1682	1598	1560	1018	-	-	-
2	-	-	3237, 3310	1675	1540	1508	1060	-	-	495
3	3450	3383	3210, 3220	1675	1550	1508	1060	-	-	490
4	3414	3383	3213, 3230	1680	1560	1508	1060	-	13851284	470
5	-	-	3240, 3215	1678	1570	1512	1069	-	-	470
6	3420	3390	3240	1676	1562	1508	1064	-	-	480
7	-	-	3210, 3290	1675	1560	1504	1060	-	-	480
8	3456	-	3250	1676	1570	1504	1053	1603, 1385	-	478
9	3400	-	3330, 3240	1659	1562	1510	1064	-	-	477, 510b
10	3400	-	3330, 3250	1643	1560	1508	1060	-	13811300	467, 500b
11	-	3360	3298, 3232	1694	1562	1551	1015	-	-	
12	-	-	3300, 3180	1696	1543	1512	1068	-	-	462
13	-	-	3305	1695	1546	1508	1061	-	13841315	465
14	-	-	3304	1685	1542	1508	1072	1600, 1381	-	466
15	-	-	3248	1679	1541	1505	1067	-	13851338	462
16	-	-	3250	1680	1538	1504	1069	-	-	486
17	-	-	3221	1685	1534	1500	1068	1610, 1385	-	460
18	-	3298 c	3170	1688	1535	1504	1068	-	-	470
19	-	33356	3225 c	1690	1537	1510	1068	-	-	474

b= υ (M- O), c = coordinated hydraxo (OH^-) group.

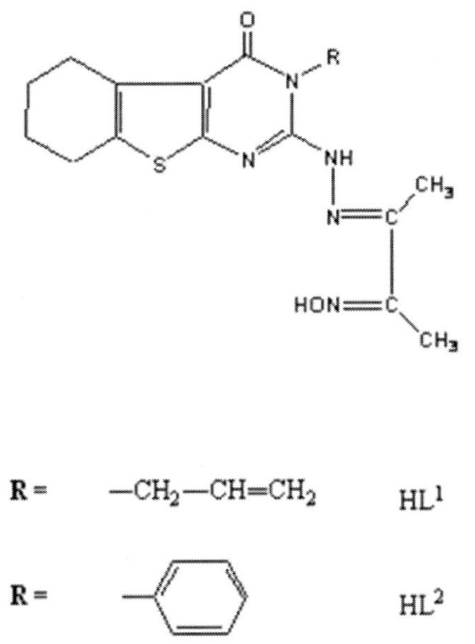

Figure 1. Chemical formulas of ligands HL^1 and HL^2.

Preparation of Metal Complexes

The metal complexes were prepared by adding a hot ethanolic solution of ligand (HL^1 or HL^2) in absence or presence of equimolar ratio of sodium hydroxide to an equimolar amount of the appropriate metal salt. The reaction mixture was heated with stirring for 2hrs. The separated complex was filtered off, washed several times with ethanol, and dried under vacuum in presence of anhydrous $CaCl_2$. The reactions of the ligands with different metal salts of Cu(II), Co(II), and Ni(II) gave complexes (1-19) as shown in table (1) and figure (2).The different metal salts used and the appropriate molar ratios are shown in table (1).

In Vitro Cytotoxicity of Pyrimidine Complexes

Newly synthesized derivatives of pyrimidine complexes were dissolved in DMSO to obtain the concentration of 1 mM. These stock solutions were stored at 4°C for further use. The toxicities of the complexes were tested in vitro against liver cell line BRL3A derived from buffalo rats [17].

Briefly, cells were harvested and suspended in RPMI-1640 medium supplemented with 10% Fetal Bovine Serum (FBS) and penicillin/ streptomycin. One hundred μl of cells at concentration of 1x 10^5/ml were titrated in 96 flat-bottomed wells plate, and then serial concentrations of each complex were added to the cells and incubated for 24 h at 37°C in 5% CO_2 incubator. The cells viability was checked by using trypan blue to detect the toxicities of complexes.

Animals Grouping

Adult male albino rats, weighing about 100-120 g, were housed at 23 ± 2°C and in daily dark/light cycle. They were maintained under standard condition and fed standard chow and water ad libitum. All experiments were carried out in accordance with protocols approved by the local experimental animal ethics committee. Animals were divided into many different groups of ten animals each. Normal group includes animals that were given only standard pellet diet and water ad libitium. CCl_4-treated group in which animals were orally received CCl_4 at a dose (2.5ml /kg, B.W) three days a week for eight weeks [18].

Corn oil-treated group in which animals were orally received 2.5 ml/kg (B.W) of corn oil three times a week for eight weeks [18]. Candesartan cilexetil/CCl_4 group in which animals were received CCl_4 (2.5ml /kg, B.W) three days a week for eight weeks, after that, they were treated with candesartan cilexetil (8 mg/kg) orally every other day for 4 weeks. In order to evaluate the potential antifibrotic activity of the pyrimidine complexes, animals were received CCl_4 (2.5ml /kg, B.W) three days weekly for eight weeks, after that, they were intraperitoneally injected with 10 μM of each complex 3 times weekly for 4 weeks. The last group of animals included normal rats injected with 100 μl of DMSO 3 times weekly for 4 weeks.

At the end of experiments, blood was collected in non-heparinized tubes and allowed to clot at room temperature. Serum samples were then obtained by centrifugation and were kept at -20°C until assayed. The liver was removed and cut into pieces. Liver homogenate were prepared by using a mechanical homogenizer (Potter-Elvehjem) in a 10-fold volume of ice-cold of 20mM tris-HCl buffer (pH 7.4). The homogenate was centrifuged at 5000 rpm for 30 minutes at 4°C to remove cell debris and nuclei. The supernatant was collected, aliquoted and kept frozen at -20°C.

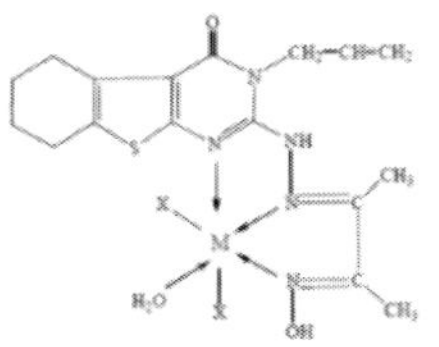

X	M	Complex	
Cl	N(II)	$[Ni(HL^1)Cl_2(H_2O)].3H_2O$	(3)
NO_3	Ni(II)	$[Ni(HL^1)(NO_3)_2(H_2O)].H_2O$	(4)
Cl	Cu(II)	$[Cu(HL^1)Cl_2(H_2O)].H_2O$	

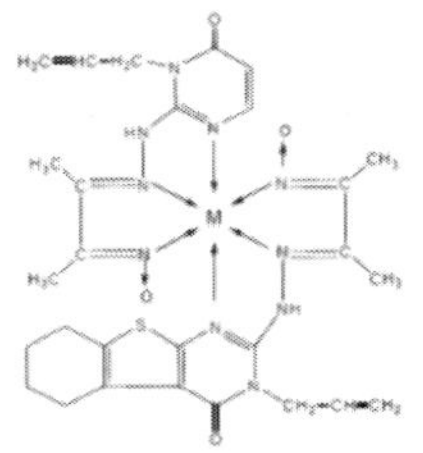

M	Complex	
Ni(II)	$[Ni(L^1)_2]$	(2)
Co(II)	$[Co(L^1)_2]$	(5)
Cu(II)	$[Cu(L^1)_2]$	(7)

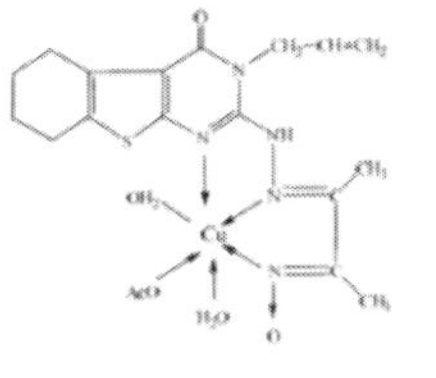

$[CuL^1(OAc)(H_2O)_2]$ (8)

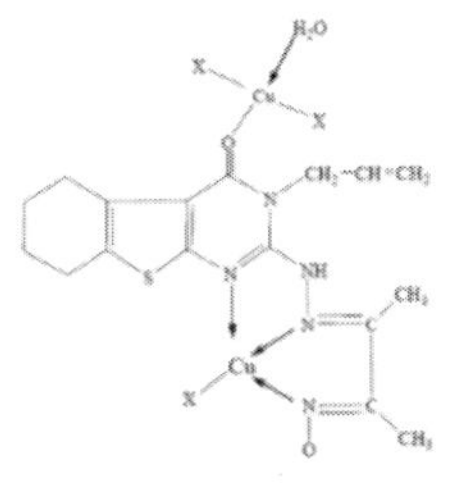

X	Complex	
Cl	$[Cu_2L^1Cl_3(H_2O)]$	(9)
NO_3	$[Cu_2L^1(NO_3)_3(H_2O)]$	(10)

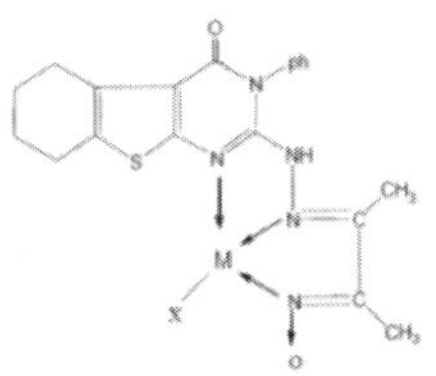

X	M	Complex	
Cl	N(II)	$[NiL^2Cl]$	(12)
NO_3	Ni(II)	$[NiL^2(NO_3)]$	(13)
OAc	Ni(II)	$[NiL^2(OAc)]$	(14)
NO_3	Co(II)	$[CoL^2(NO_3)]$	(15)
ClOAc	Co(II)	$[CoL^2(OAc)]$	(17)
OH	Cu(II)	$[CuL^2(OH)].H_2O$	(18)[17]

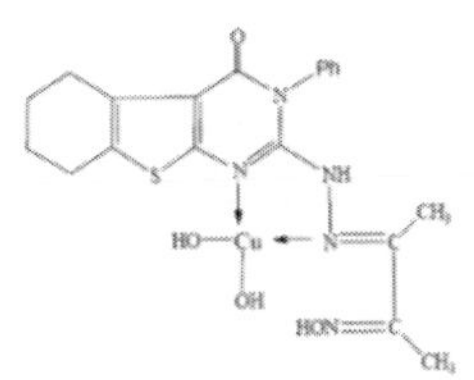

$CuL^2(OH)_2$ (19)

Figure 2. Chemical formulas of metal complexes.

Liver Function Investigations

In serum of all animals, activities of aspartate aminotransferase (AST), alanine aminotransferase (ALT), gamma glutamyl transferase (γGT) and the level of albumin were detected by a Beckman automatic analyzer.

Determination of Hepatic Hydroxyproline

Hydroxyproline (HPO) content in liver was determined by the method of [19]. The serum samples were hydrolyzed in 6 N HCl (final conc.) in sealed tubes at 110°C for 16 h. The hydrolyzed samples were evaporated to dryness in a boiling water bath to remove acid, and the residue was dissolved in distilled water and made up to a known volume. It was decolorized with activated charcoal and filtered through Whatman filter paper. For the determination of hydroxyproline, 1 ml of the clear filtrate was mixed with 1 ml of freshly prepared chloramine-T solution and allowed to stand for 20 min.

The samples were further mixed with 1 ml of 3.15 mol/l perchloric acid and waited for 5 min. Finally, 1ml of freshly prepared *p*-dimethylamino benzaldehyde was added, mixed well, and placed in a water bath at 60°C for 20 min. The absorbance of the solution was determined by using a spectrophotometer at 560 nm.

Estimation of Oxidative Status Markers

The hepatic activity of superoxide dismutase was estimated according to the procedure of Nishikimi et al. [20] and by following the manufacturer's procedure (Biodiagnostics, Egypt). Catalase activity in liver homogenate was estimated by the method of Gaetani et al. [21] and according to the manufacturer's procedure (Biodiagnostics, Egypt). Level of reduced glutathione in liver homogenate was estimated by using the method mentioned by Prins and Loose [22]. Level of lipid peroxidation in liver homogenate was estimated colorimetrically by measuring malondialdehyde (MDA) using the method of Ohkawa et al. [23] and by following the manufacturer's procedure (Biodiagnostics, Egypt).

Detection of Serum Levels of TNF-A and Fibronectin

The serum levels of TNF-α and fibronectin were detected by using enzyme-linked immunosorbent assay (ELISA) and following the manufacture's procedure of Immunotech Company (France) and ALPCO Diagnostics Company (USA); respectively.

Statistical Analysis

All data were expressed as Mean values ± SD. *t* test was used for comparison between groups. *P* values less than 0.05 were regarded as statistically significance.

Results and Discussion

Elucidation of Chemical Structures of Ligands and Their Metal Complexes

The prepared metal complexes are stable at room temperature, non-hygroscopic and partly soluble in most organic solvents and water, but completely soluble in dimethylformamide (DMF) or dimethylsulphoxide (DMSO). Elemental analyses, physical data (Table 1) and spectral data (Table 2), are consistent with the proposed structures as shown in figure (2). The elemental analyses confirmed that the complexes, 3, 4, 6, 8, 12-18 and 19 are formed in 1:1 molar ratio between the metal ion and ligand whereas, complexes 2, 5, and 7 are formed in 1:2 molar ratio between the metal ion and the ligand. Complexes 9 and 10 are binuclear complexes.

The Molar Conductance

The molar conductance values of the metal complexes in DMF (10^{-3}M) are listed in table (1). The values of molar conductance show that all complexes are non-electrolytes [24, 25]. This confirms that, the anions in all these complexes are directly attached to the metal ion.

Infrared Spectra

The most diagnostic infrared spectral bands of the ligands; HL^1 and HL^2 are depicted in table (2). The spectra of ligands HL^1 and HL^2 show bands at 3333-3360, 3310-3298, and 3237-32323 cm^{-1}, the first band assigned to ν(O-H)oxime and the later bands assigned to ν(N-H). The spectra also display bands at 1694-1682, 1598-1562, and 1560-1551 cm^{-1}, assigned to ν(C=O), ν(C=N)imine, and ν(C=N)oxime; respectively.

In the infrared spectra of ligands, the bands characteristic to pyridmidine ring appear at 1578 and 632 cm^{-1}. The N-O band in the spectra of the two ligands appears at 1018-1015 cm^{-1}. These values are in accord with the previously reported hydrazone and oxime derivatives [26-29]. The mode of bonding of the ligands, HL^1 and HL^2 in the metal complexes have been deduced from the IR spectra. The most important diagnostic spectral bands of the ligands and their metal complexes are depicted in table (2).

Results shows that the bands corresponding to ν(C=N)imine and ν(C=N)oxime in the spectra of all complexes except, complex 19 are shifted to lower wave-numbers compared to those of the free ligands, as a result of their coordination to the central metal ion. The spectra of all complexes also reveal that the bands corresponding to the pyrimidine ring are perturbed compared to those of the free ligands, indicating that the pyrimidine nitrogen atom is involved in coordination. The stretching vibration band ascribed to the carbonyl group appears at the same position as those of the free ligands except, in the spectra of binuclear complexes 9 and 10 is shifted to a lower wave-number compared to those of the free ligands, indicating that the carbonyl oxygen atom does not participate in coordination in all complexes except, complexes 9 and 10 it is participated in coordination to the metal ion. The infrared spectra of all complexes except, the complexes 3, 4 and 6 show that the band characteristics to ν(O-H)oxime disappears upon complexation, indicating that the hydroxyl group ionized upon coordination. This could be taken as evidence that the ligands in these complexes behave as a mono-basic ligand. Whilst, this band remains in the spectra of complexes 3, 4 and 6 revels that it does not ionize in these complexes ,indicating that the ligand behaves as a neutral ones. The infrared spectra of all metal complexes also revealed that the bands ascribed to ν(N-H) appear at the same positions or slightly shifted compared to those of the free ligands, indicating that it does not involved in coordination.

The band ascribed to ν(N-O)oxime in the spectra of metal complexes is shifted to a higher wave-number relative to that of the free ligand, indicating that the oxime group is coordinated via the nitrogen atom. The infrared spectrum of complex 19 show that the ν(C=N)oxime remains at the same position as that of the ligand in addition the ν(O-H)oxime does not disappear upon complexation.

The above arguments indicate that the ligands behave as a mono-basic tridentate ones, coordinated via the imino nitrogen, oximino nitrogen and pyrimidine nitrogen atoms as in case of complexes 2, 5 7, 8, 9, 10, and 12-18.

Ligand HL^1 also behaves as a neutral tridentate ones as in complexes 3, 4 and 6. Moreover, HL^2 behaves in complex 19 as a neutral didentate ones, the ligand coordinated through the imino nitrogen and pyrimidine nitrogen atoms. In binuclear complexes 9 and 10 the ligand binds two copper(II) ions, one ion is coordinated via the imino, oximino and pyrimidino nitrogen atoms, the second copper(II) ion is coordinated via the carbonyl oxygen atom. The rest of coordination number is satisfied by the appropriate anion and water as shown in figure (2).

The spectra of nitrate complexes 4, 10, 13 and 15 show a strong band at 1385-1381 cm^{-1}, as well as a weak band at 1338- 1284 cm^{-1}, was assigned to unidentate nitrate [30]. The spectra of the acetate complexes 8, 14 and 17 show two absorption bands at 1610-1600 and 1385-1381cm^{-1}, assigned to υ(C=O) and υ(C-O) , respectively. The difference between these two bands is near 220 cm^{-1}, indicates that the acetate in this complex coordinates to the metal ion as a unidentate ligand [31].

The IR spectra of the hydrated complexes display a broad band at 3456-3400 cm^{-1}, assigned to ν(OH) of water molecule [32]. The IR spectra of all complexes display a new band at 495-460 cm^{-1}, assigned to ν(M-N) [27- 29]. The spectra of metal complexes 9 and 10, show a new band 510-500 cm^{-1}, assigned to υ(M-O) [33-35].

Electronic Spectra and Magnetic Moment

The electronic spectra of the copper(II) complexes, 6, 7 and 8 show two bands at 860 – 710 nm and 680 – 510 nm, assigned to $^2B_1g \rightarrow {}^2Eg$ and $^2B_1g \rightarrow {}^2A_1g$ respectively, suggesting a distorted tetragonal geometry [34-36]. However, the electronic spectra of copper(II) complexes 9, 10, 18 and 19 exhibit a broad band at 640-520 nm, which suggesting a square planar geometry around the copper(II) ion [31]. The nickel(II) complexes 2, 3 and 4 display electronic spectra typical of an octahedral system. The spectra of these complexes show three bands at 780 nm, 600 nm and 460 nm which may be assigned to the transitions $^3A_2g(F) \rightarrow {}^3T_1g(P)(\upsilon_3)$, $^3A_2g(F) \rightarrow {}^3T_1g(F)(\upsilon_2)$ and $^3A_2g(F) \rightarrow {}^3T_2g(F)$ (υ_1) respectively, in octahedral symmetry [32- 34].The υ_2/υ_1 ratio is near 1.3, which is less than the usual range of 1.5 – 1.75, indicating distorted octahedral nickel(II) complexes [41, 42]. The electronic spectra of cobalt(II) complexes 12, 13, and 14 display the ν_3 band as a strong multiple band around 650 nm, suggesting a tetrahedral geometry of these complexes.. However, complex 5 displays the ν_3 band as a weak band at 780 nm, suggesting an octahedral arrangement around cobalt(II) ion. The magnetic moment values per one metal ion measured at room temperature were listed in table (3). The values indicate that all metal ions in all metal complexes display values for spin only values, indicating no any type of magnetic interactions between the metal ions. The molecular formulas of the obtained complexes were shown in figure (2).

Table 3. Solid state electronic spectral bands for metal complexes

No.	Compounds	d-d bands (nm)	μeff (B.M.)/metal ion
1	HL^1		
2	$[Ni(L^1)_2]$	780, 600, 460	2.90
3	$[Ni(HL^1)Cl_2(H_2O)].3H_2O$		2.83
4	$[Ni(HL^1)(NO_3)_2(H_2O)].H_2O$		2.95
5	$[Co(L^1)_2]$	780, 520	4.93
6	$[Cu(HL^1)Cl_2(H_2O)].H_2O$	680(br), 480(br)	1.89
7	$[Cu(L^1)_2]$	750, 510, 440(sh)	1.93
8	$[CuL^1(OAc)(H_2O)_2]$	710, 510, 440	1.79
9	$[Cu_2L^1Cl_3]$	520(br)	1.98
10	$[Cu_2L^1(NO_3)_3(H_2O)]$	640, 470(br)	1.88
11	HL^2		
12	$[NiL^2Cl]$	560, 475(br)	2.93
13	$[NiL^2 (NO_3)]$	600, 390	3.20
14	$[NiL^2(OAc)]$	600, 460	3.25
15	$[Co(L^2)(NO_3)]$	650, 520	4.30
16	$[Co(L^2) Cl]$	623, 512	4.41
17	$[CoL^2 (OAc)]$	510, 480	4..23
18	$[Cu(L^2)(OH)] .H_2O$	585(br), 490, 430(sh)	1.93
19	$[CuL^2(OH)_2]$	600, 480	1.85

Effect of CCl_4

Eight weeks administration of CCl_4 to rats resulted in a significant rise in liver fibrosis as compared to the control groups of animals. This rise in liver fibrosis was concluded by the significant increase in liver enzymes and the level of hepatic hydroxproline (HPO). Table (4) shows the values for the activities of the serum indicator enzymes and the markers of hepatic fibrosis. It is evident that CCl_4 produced a statistically marked increase ($P>0.05$) in the activities of serum ALT, AST and γ-GT as compared to the control and sham groups of animals. Table (4) showed that treatment with CCl_4 for 8 weeks led to a significant elevation in hepatic HPO as compared to the control and sham groups. The results also showed that CCl_4 reduced the capacity of liver to produce albumin, where the level of albumin was statistically diminished compared to the normal and sham groups of animals.

Table 4. Effect of different derivatives of pyrimidine complexes on liver enzymes, albumin and hepatic hydroxyproline

Treatment	ALT (U/L)	AST (U/L)	Alb (mg/dl)	γGT (U/L)	HPO (mg/g tissue)
Normal	43.2 ±3.4	158.6±11.2	3.6±0.5	12.0±1.6	0.33±0.02
CCl_4	102.4±8.9*	218.5±16.7*	1.0±0.5*	18.2±1.7*	1.12±0.03*
CCl_4/CC	60.1±5.2**	182.0±16.8**	2.9±0.6**	13.5±2.3**	0.49±0.01**
Corn oil	44.7±9.8	160.1±9.5	3.4±1.1	11.4±4.2	0.26±0.02
DMSO	45.6±5	166.5±15.4	3.5±1	12.3±2.2	0.39±0.1
(1), HL^1	82.2±6.8**	181.3±14.5**	2.2±0.4**	15.2±1.9	0.67±0.02**
(5), HL^1+$Co(NO_3)_2$ (1:1)	68.7±5.7**	172.3±14.5**	2.8±0.4**	13.6±1.7**	0.56±0.02**
(2), HL^1+$NiCl_2$ (1:1:1)	77.2±7.4**	177.2±16.7**	2.2±0.3**	14.2±2.4**	0.79±0.04**
(4), HL^1+$Ni(NO_3)_2$ (1:1)	81.1±8.3**	178.0±7**	2.3±0.7**	14.1±5.2**	0.92±0.05**
(11), HL^2	76.6±8.9**	180.5±6.4**	3.0±0.3**	13.1±1.6**	0.50±0.06**
(16), HL^2+$CoCl_2$ (1:1:1)	55.0±6.8**	171.3±7.6**	3.0±0.6**	12.7±4.1**	0.42±0.05**
(15), HL^2+$Co(NO_3)_2$ (1:1)	59.9±9.7**	173.6±7.9**	2.9±0.4**	12.7±2.1**	0.40±0.05**
(12), HL^2+$NiCl_2$ (1:1:1)	71.0±6.5**	163.8±6.9**	3.0±0.6**	11.2±2.1**	0.39±0.06**
(13), HL^2+$Ni(NO_3)_2$ (1:1)	61.7±7.7**	166.3±9.3**	2.7±0.7**	10.8±2.1**	0.38±0.05**

(*) Significant as compared to the normal or sham group.
(**) Significant when compared to the CCl_4-treated group.

Hepatic antioxidation data, shown in table (5), indicated that lipid peroxidation was increased, while, SOD and catalase activities were decreased in CCl_4-treated rats significantly, as compared to those of control and sham groups. Also, the GSH content in CCl_4 treated group was significantly lowered as compared to that of normal and sham groups. Furthermore, results indicated that the level of TNF-α and fibronectin (Fbn) have significantly increased in the serum of rats treated with CCl_4 for 8 weeks as compared to those of normal and sham groups as shown in table (6).

Table 5. Effect of different derivatives of pyrimidine complexes on the hepatic antioxidant status

Treatment	CAT (μmol/Sec/g)	SOD (U/g)	GSH (mg/g)	LPO (nmol/g)
Normal	6.2±2.2	162.5±18.3	4.3±1.1	135.3±16.7
CCl_4	1.3±0.4*	99.2±15.3*	1.1±0.4*	252.8±23.9*
CCl_4/CC	5.2±2.1**	152.8±28.1**	3.5±1.3**	142.5±21.8**
Corn oil	6.0±2.3	160.1±23.1	4.2±1.2	134.3±13.9
DMSO	6.2±1.9	159.6±19.4	3.9±1.1	133.2±14.2
(1), HL^1	3.9±1.1**	134.0±19.8**	2.8±0.8**	215.0±21.3**
(5), HL^1+$Co(NO_3)_2$ (1:1)	4.9±1.3**	148.8±21.9**	2.6±0.9**	152.6±15.3**
(2), HL^1+$NiCl_2$ (1:1:1)	4.4±1.0**	142.2±19.9**	3.7±1.5**	172.6±18.9**
(4), HL^1+$Ni(NO_3)_2$ (1:1)	4.8±1.3**	150.6±16.7**	4.0±1.6**	157.0±19.3**
(11), HL^2	4.6±1.7**	153.0±14.3**	3.1±1.1**	195.2±18.3**
(16), HL^2+$CoCl_2$ (1:1:1)	5.1±1.3**	150.8±10.4**	4.1±1.2**	160.4±19.3**
(15), HL^2+$Co(NO_3)_2$ (1:1)	6.0±1.6**	156.8±13.2**	4.0±1.3**	153.4±18.4**
(12), HL^2+$NiCl_2$ (1:1:1)	5.8±1.1**	152.6±13.9**	4.2±1.7**	148.6±17.9**
(13), HL^2+$Ni(NO_3)_2$ (1:1)	6.1±2.1**	153.4±13.2**	4.2±1.9**	145.7±21.0**

(*) Significant as compared to the normal or sham group.
(**) Significant when compared to the CCl_4-treated group.

Effect of Candesartan Cilexetil (CC)

Treatment of CCl_4-intoxicated rats with Candesartan cilexetil (CC) for 4 weeks showed a modulatory action on the liver fibrosis indicators. As shown in table (4), administration of CC to the liver fibrotic rats resulted in a significant amelioration in the level of hepatic enzymes as compared to those of CCl_4-induced liver fibrosis group. Meanwhile, hepatic hydroxyproline has significantly decreased after administration of CC as compared to that of CCl_4 treated animals.

Table (5) shows the effect of CC on the level of antioxidant status after induction of liver fibrosis with CCl_4. Results indicated that CC ameliorated the antioxidant status, where the level of LPO was significantly decreased compared to that of CCl_4-intosicated rats. On the other hand, the level of GSH and SOD and catalase activities were increased significantly as compared to those of CCl_4-treated animals.

Moreover, levels of TNF-α and fibronectin in serum of rats suffered from liver fibrosis were significantly decreased after administration of CC for 4 weeks as compared to those of CCl_4-treated animals (Table 6).

Table 6. Levels of serum TNF-α and fibronectin (Fbn) in different groups of the study

Treatment	Fbn (µmol/Sec/g)	TNF-α (ng/L)
Normal	32.7±1.2	14.8±1.3
CCl_4	85.7±6.0*	20.7±1.4*
CCl_4/CC	42.3±5.0**	15.5±1.1**
Corn oil	32.0±1.4	13.2±1.2
DMSO	35.7±2.1	15.6±0.9
(1), HL^1	74.7±4.9**	16.7±1.1**
(5), HL^1+$Co(NO_3)_2$ (1:1)	67.0±12.8**	17.2±0.7**
(2), HL^1+$NiCl_2$ (1:1:1)	60.7±8.4**	16.4±0.8**
(4), HL^1+$Ni(NO_3)_2$ (1:1)	75.7±5.2**	16.1±0.3**
(11), HL^2	64.6±6.7**	19.9±0.6
(16), HL^2+$CoCl_2$ (1:1:1)	36.0±3.5**	15.3±1.1**
(15), HL^2+$Co(NO_3)_2$ (1:1)	51.0±15.7**	16.7±0.6**
(12), HL^2+$NiCl_2$ (1:1:1)	65.0±15.1**	16.8±1.0**
(13)HL^2+$Ni(NO_3)_2$ (1:1)	79.0±2.0**	17.2±1.1**

(*) Significant as compared to the normal or sham group.
(**) Significant when compared to the CCl_4-treated group.

Effect of Different Derivatives of Pyrimidine Complexes on Liver Fibrosis

Chloride and nitrate derivatives of pyrimidine complexes were synthesized and dissolved in DMSO solution. All complexes were dissolved in DMSO solution and checked for their antifibrotic activity in liver fibrosis model induced in rats with CCl_4 intoxication. The cytotoxicity of these complexes were investigated against normal liver cell line BRL3A and the LC_{50} was determined for all tested complexes.

Results showed that the LC_{50} for most complexes was around 80 μM (Data not shown). The possible antifibrotic potential activity of these newly synthesized complexes have been tested in a final concentration of 10μM by injection intraperitoneally in rats intoxicated with CCl_4.

We observed no death was occurred among animals treated with these complexes, in addition, the livers obtained from animals injected with the tested complexes seemed mostly as the livers obtained from the normal control group.

Results indicated that most of tested complexes exhibited an ameliorated effect on the liver fibrosis. The serum levels of liver enzymes were significantly decreased after treatment with pyrimidine complexes as compared to those of CCl_4-intoxicated animals (Table 4). Administration of most of pyrimidine complexes to rats with liver fibrosis resulted in a significant augmentation in the hepatic antioxidant capacity as compared to those of CCl_4-trated group (Table 5). Moreover, administration of most of pyrimidine complexes to the CCl_4-intoxicated rats led to significant depletion in the fibrotic parameters, hepatic hydroxyproline (HPO), TNF-α and fibronectin (Fbn), as compared to those of CCl_4-treated animals.

Effect of Ligands on the Antifibrotic Activity of the Complexes

We examined the effect of ligands on the antifibrotic activity of these complexes. Three distinguished indicators for liver fibrosis have been tested, hepatic hydroxyproline (HPO), TNF-α and fibronectin (Fbn); therefore the percentage of reduction of these indicators after administration of these complexes to CCl_4-intoxicated rats were shown in figures (1 and 2).

Results shown in figure (1 A and B) revealed that cobalt derivatives of ligand HL^2 complexes have potent antifibrotic activities rather than complexes of ligand HL^1. The percentage of antifibrotic activity, referred to CCl_4-treated group, was significantly increased in complexes having ligand HL^2 of cobalt complexes. In case of TNF-α, results indicated that both ligands-containing complexes had the same effect on the percentage of inhibition of TNF-α. Complexes of nickel with ligand HL^1 were compared with those having lignad HL^2 (Figures 2A and B). Like cobalt complexes of ligand HL^2, nickel complexes of ligand HL^2 exhibited antifibrotic activity higher than those with ligand HL^1. Also, we noticed that TNF level was mostly resistant to the effect of ligand HL^2.

Effect of Metal Ion Derivative on the Antifibrotic Activity of the Complexes

The percentage of antifibrotic activity of different metal complexes of ligand HL^1were compared in regarding to that of the free ligand HL^1 (Figures 3A and B). Results indicated that the antifibrotic activities of different metal complexes of ligand HL^1 were significantly increased as compared to that of free ligand HL^1. Cobalt(II) complexes exhibited potent antifibrotic activity higher than that of nickel(II).

Figures (4A and B) show the percentage of antifibrotic activities of different metal complexes of ligand HL^2 in parallel with that of the free ligand HL^2. Results revealed that the antifibrotic activity of ligand HL^2 was increased significantly when coordinated with different metal ions. Like complexes of ligand HL^1, ligand HL^2 exhibited high antifibrotic potency when coordinated with cobalt(II) ion.

Hepatic fibrosis can lead to the development of hepatic cirrhosis and hepatocellular carcinoma [42]. The development of approaches to prevent hepatic fibrosis is therefore important although no available drugs can prevent or cure liver fibrosis [1]. The present study was designated to investigate the antifibrotic activity of newly synthesized pyrimidine complexes in liver fibrosis in rat induced by CCl_4.

Previous studies reported the leishmanicidal [44], antimicrobial and antibacterial [45] activity of the pyrimidine complexes. To our knowledge, this is the first trial to investigate the antifibrotic activity of these complexes in rat model induced by CCl_4. The present study demonstrated a significant elevation in oxidative stress markers after 3 times treatment with CCl_4 for 8 weeks. Oxidative stress is due to the liberation of reactive oxygen species (ROS) that can lead to liver damage.

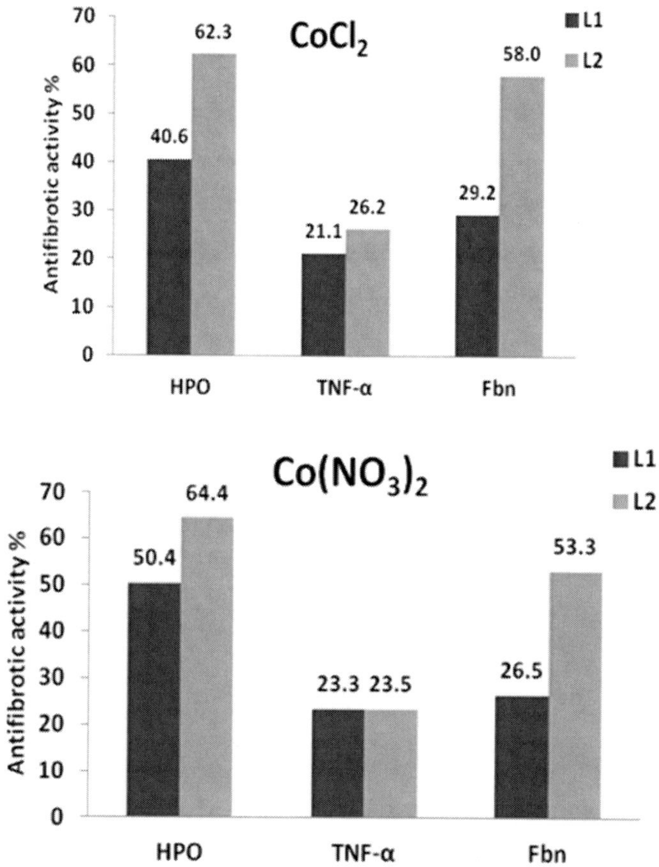

Figure 3A and B. The effect of ligand on the antifibrotic activity of Cobalt derivatives of pyrmidine complexes.

Oxidative damage occurred when free radicals produced within an organism were not completely eliminated by the appropriate endogenous antioxidant systems. Thus, the present results indicated disruption of the proxidant/antioxidant balance in the liver and supported earlier work published by [46]. Oxygen radicals lead to liver damage ending with liver fibrosis. The body has an effective defense mechanism to prevent and neutralize the free radical induced damage. This is accomplished by a set of endogenous antioxidant enzymes such as superoxide dismutase, catalase and GSH system.

Peroxidation of lipids can dramatically change the properties of biological membranes, resulting in severe cell damage and could play a significant role in the pathogenesis of disease. It has showed that lipid peroxidation, free-radical mediated process, and certain lipid peroxidation products induce genetic over expression of fibrogenic cytokines and increase the synthesis of collagen. Free radicals and MDA can stimulate the synthesis of collagen and initiate liver fibrosis [47].

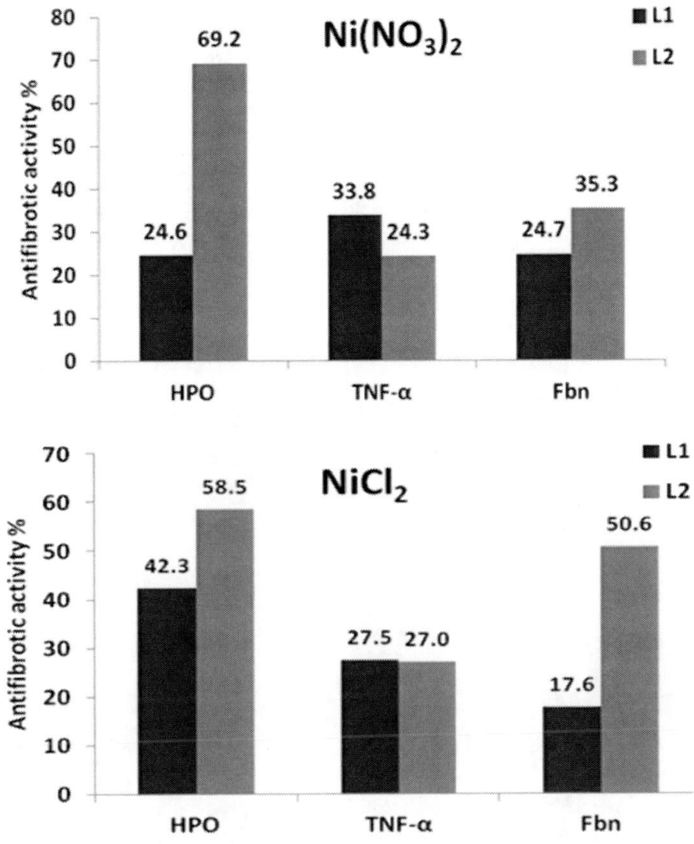

Figure 4A and B. The effect of ligand on the antifibrotic activity of Nickel derivatives of pyrmidine complexes.

Elevation in lipid peroxides and depletion in the activities of SOD and catalase mean deterioration in the hepatoprotective system resulting in an inflammation that is mediated by many cytokins such as TNF-α.

TNF-α was originally identified as a circulating factor that resulted in remarkable hemorrhage and necrosis of tumors when administered to tumor-bearing mice. It has been implicated in a number of liver diseases and is an important mediator of many physiological conditions [48]. In several animal models of immune-mediated liver injury or hepatotoxin sensitization, TNF-α administration could lead to hepatocyte apoptosis and liver failure. Besides its modulation effect on extracellular matrix (ECM) production, TNF-α has been considered a mediator of cell injuries in liver caused by alcoholism, reperfusion, primary graft nonfunctional, graft rejection and endotoxic insult [49]. As inflammation occurs in the liver, the formation of fibrosis is induced which leads to a significant increase in the level of hydroxproline (HPO).

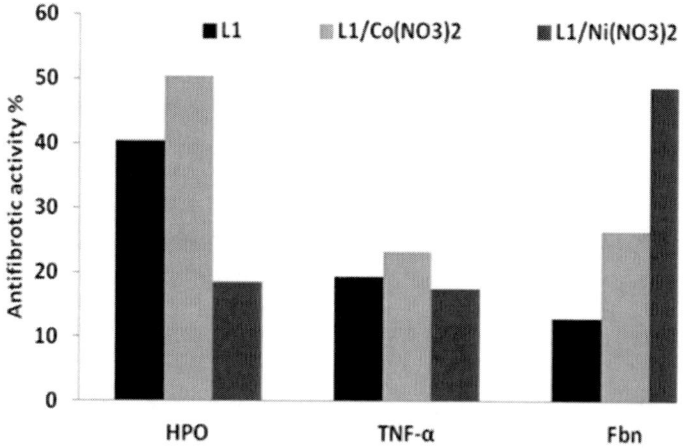

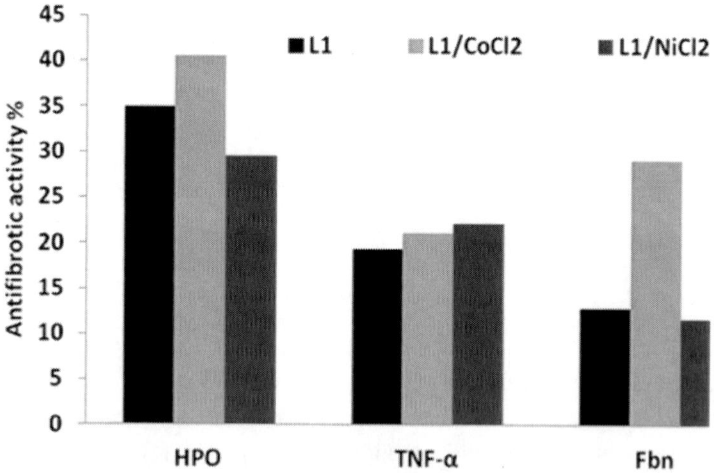

Figure 5A and B. The effect of metal derivatives on the antifibrotic activity of pyrmidine complexes.

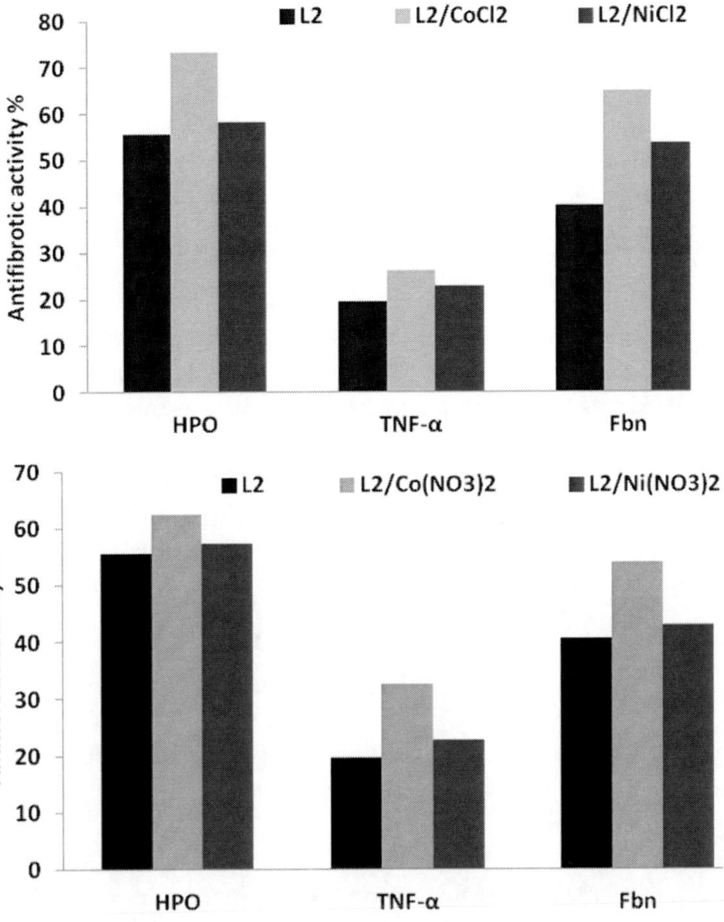

Figure 6A and B. The effect of metal derivatives on the antifibrotic activity of pyrmidine complexes.

Recent study reported that candesartan cilexetil (CC) showed marked significant decrease in liver fibrosis indices, as evidenced by significant reduction of TNF-α and lipid peroxides. They also revealed that CC may further inhibit liver fibrogenesis in CCl_4-treated rats. It proved that CC significantly suppressed injury-induced TGF-β1 expression in liver and attenuated fibrogenesis, as evaluated by hydroxproline content in liver [50].

Our results indicated that the pyrimidine complexes mostly exhibited antioxidant activity as they were able to increase the antioxidant capacity in the liver and to prevent the formation of lipid peroxides. Therefore, we recommend that the antifibrotic activity of these complexes may attribute to their antioxidant activity. As mentioned above that the fibrogenesis as a result of CCl_4 intoxication is due to the liberation of ROS and lipid peroxides in hepatic tissue. So, the scavenging of free radical might lead to amelioration of antioxidant capacity and to recover fibrosis by induction the hepatocytes to compensate the damaged cells. Furthermore, these complexes exhibited varied antifibrotic effect depending on both type of ligand and type of metal ion coordinated in the complex structure. Our result indicated that the antifibrotic activity of ligand HL^2 complexes was more potent than complexes of ligand HL^1. By analysis the chemical structure of both ligands, we may attribute the activity of ligand HL^2 to the presence of phenyl group instead of propylene side chain in ligand HL^1. The presence of phenyl group acts as electron withdrawal due to the resonance of benzene ring and this may increase the ROS-scavenging activity of ligand HL^2 and hence its antifibrotic activity.

Also, our results indicate that the type of metal ion affected the antifibrotic potential activity of these complexes. Results proved that cobalt complexes exhibited higher antifibrotic activity rather than nickel metal complexes; the order of potency was as follows: Co > Ni.

Conclusion

In the present study, we screened panels of newly prepared complexes of pyrimidine derivatives with different metal ions for their anti liver fibrosis activity induced by CCl_4 in rats. The study indicated that most of these complexes exhibited very promising activities against liver fibrosis in vivo. The mechanism of action of these complexes is suggested to be through their antioxidant potential activity which leads to scavenging the ROS resulting in augmentation of hepatic antioxidant capacity. The activities of these complexes are related to their structures where complexes of ligand HL^2 exhibited potent antifibrotic activity rather than that of ligand HL^1 complexes. Meanwhile, cobalt complexes exhibited higher antifibrotic activity rather than nickel metal complexes. We recommend conducting a further intensive study to investigate the molecular mechanism of these complexes.

Acknowledgment

Thanks are due to Scientific Research Deanship, Qassim University, KSA, for supporting this work.

REFERENCES

[1] Bataller R., Brenner D. *J Clin Invest.*, 115, (2005) 209-218.
[2] Aziz-Seible R.S., Benita L.M., Kusum K.K. Carol A.C., *World J Hepatol.*, 27, (2), (2011) 45-55.
[3] Zangar R.C., Benson J.M., Burnett V.L., Springer D.L. *Chem. Biol. Interact.*, 125(3), (2000) 233-243.
[4] Weber L.W., Boll M., Stampfl A. *Crit. Rev. Toxicol.*, 33, (2003) 105-136.
[5] Stenhoff H., Lagerstrom P.O., Andersen C., *Chromatogr J.* B 731, (1999) 411-417.
[6] Paizis G., Gilbert R.E., Cooper M.E., Murthi P., Schembri J.M., Wu L.L.. *J Hepatol.,* 35, (2001) 376-385.
[7] Hayashi S., Itoh A., Isoda K., Kondoh M., Kawase M., Yagi K. *Eur J Pharmacol.*, 580, (2008) 380–384.
[8] Sakurai H, Kojima Y, Yoshikawa Y, Kawabe K, Yasui H. *Coord Chem Rev.*, 226, (2002) 187-198.
[9] Sadler PJ, Li H, Sun H. *Coord Chem Rev.*, 185, (1999) 689-709.
[10] Bhupinder S. S. *J Pharm Educ Res.,* 1(1), (2010) 1-20.
[11] Varadinova T., Shishkov S., Panteva M., Bontchev P. *Metal-Based Drugs*. 3(3), (1996) 149-154
[12] Ali H, van Lier JE. *Chem Rev.*, 99, (1999) 2379-2450.
[13] Louie AY, Meade TJ. *Chem Rev* 99, (1999) 2711-2734.
[14] Yoshiji H., Kurijoma S., Yoshii J., Ikenada Y., Noguchi R., Nakatani T. *Hepatol.*, 34, (2001) 745-750.
[15] Gewald K., Schinke E., Bottcher H. Ber, 99, (1966) 94.
[16] Lafi A. M.Sc. Thesis (Chemistry), College of Science, Qassim University (2010).
[17] Nissley S.P., Short P.A., Rechler M.M., Podskalny J.M., Coon H.G.. *Cell.* 11(2), (1977) 441-6.
[18] Fischer-Nielsen A., Poulsen H.E., Hansen B.A., Hage E., Keiding S. *J. hepatol.*, 12, (1991) 112-117.
[19] Sagar P.T., Hitesh N.J., Savita D.P., Umesh M.U.. *Bangl J Pharmacol.*, 4, (2009) 126-130.
[20] Nishikimi M., Rao N.A., Yog K.. *Bioch. Biophys. Res. Commun.*, 46, (1972) 849-851.
[21] Gaetani G. F., Ferraris A. M., Rolfo M., Mangerini R., Arena S., Kirkman H. N. *Blood.* 87, (1996) 1595-1599.

[22] Prins H.K., Loose J.A. Glutathione. In: *Biochemical Methods in Red Cell Genetics*. Edited Academic Press. NY., London, (1969) 126-129.
[23] Ohkawa H., Wakatsuki A., Kaneada C. *Anal Biochem*., 95, (1982) 351-358.
[24] Ainscouhg A.W., Plowman R.A. *Aust J Chem*. 23, (1970) 699.
[25] Devoto G., Massocessi M., Pinna R., Ponticelli G.. *Spectrochim. Acta* A., 38, (1982) 725.
[26] Reddy K. H., Prasod N. B. *Talanta,* 59, (2003) 425-433.
[27] Kaminsky W., Jasonsli J.P., Woudenberg R., Goldberg K.I., West D.J. *Mol. Struct*., 608, (2002) 135.
[28] Gup R., Russ J. *Coord. Chem*., 32, (2005) 1.
[29] Gup R., Nakamoto K. "*Infrared and Raman Spectra of Inorganic and Coordination Compounds*" 3rd Edn. John Wileyand Sons, New York, (1970) 232-233.
[30] El-Tabl H. M., El-Saied F. A. *Synth. React. Inorg. Met-Org. Nano-Met. Chem*., 35, (2005) 243.
[31] Gupta L.K., Bansal U., Chandra S. *Spectrochim. Acta* A., 66, (2007) 972.
[32] Maurya R.C., Rajput S. *J Mol Struct*., 833, (2007) 133.
[33] Abd El-Wahab Z.H., Mashaly M.M., Salman A.A., El-Shetary B.A., Faheim A.A. *Spectrochim. Acta* A., 20, (2004) 2860.
[34] Sallam S.A., Orabi A.S., El- Shetary B.A., Lentz A. *Transition Met. Chem*., 27, (2002) 447.
[35] Tossidis A., Bolos C.A. *Inorg Chim Acta*. 112, (1986) 93.
[36] Ainscough E.W., Brodie A.M., Dobbs A.J., Ranford J.D., Waters J.M. *Inorg Chim Acta.* 267, (1998) 27.
[37] Lever A.B.P. *Inorganic Electronic Spectroscopy*, second edition, Elsevier Science Publishing Company, Amsterdam, (1984).
[38] Gubasi K.B., S.A. Patil, R.S. Vadvavi, R.V. Shenoy, M. Nethaji. *Transition Met. Chem.,* 31, (2006) 586.
[39] Thakkar N.V., Bootwala S.Z. *Ind J Chem*., 34, (1995) 370.
[40] Sathyanarayana D.N. Electronic Absorption Spectroscopy and Related Techniques, Orient Longman Limted © Universities Press (India) Limited (2001).
[41] El-Tabl A.S., El-Enein S.A. *J. Coord. Chem*., 57, (2004) 281.
[42] Chiumia G.C., Phillips D.J., Rae A.D. *Inorg Chim Acta*. 238, (1995) 197.
[43] Nissen N.N., Martin P. *J Clin Gastroenterol*., 35, (2002) 79-85.

[44] Boutaleb-Charki S, Marín C, Maldonado CR, Rosales MJ, Urbano J, Guitierrez-Sánchez R, Quirós M, Salas JM, Sánchez-Moreno M. *Drug Metab Lett.*, 3(1), (2009) 35-44.

[45] Dhankar RP., Rahatgaonkar AM., Chorghade MS., Tiwari A. Spectrochim Acta A Mol *Biomol Spectr.*, 93, (2012) 348-53.

[46] Bhandarkar M.R., Khan A. *J Ethnopharmacol.*, 91, (2004) 61-64.

[47] Jun-Wang Xu., Jun Gong, Xin-Ming Chang, Jin-Yan Luo, Lei Dong, Zhi-Ming Hao, Ai Jia, Gui-Ping Xu. *World J Gastroenterol.*, 8(5), (2002) 883-887.

[48] Roberts R., James N., Cosulich S. *Hepatol.*, 31, (2000) 420-427.

[49] Crespo J., Cayon A., Fernandez-Gil P., Hernandez-Guerra M., Mayorga M., Dominguez-Diez A., Fernandez-Escalante J.C., Pons-Romero F. *Hepatolo.*, 34, (2001) 1158-1163.

[50] Inas E., Abdulaziz H. *Bull Alex Fac Med.*, 42(1), (2006) 23-29.

INDEX

C

D

E

F

G

N

O

P

Q

R

S

T

U

V